Pipeline Coatings

Y. Frank Cheng, FNACE
Richard Norsworthy, NACE Corrosion Specialist

NACE INTERNATIONAL

NACE International
The Worldwide Corrosion Authority

Printed in the United States of America

ISBN: 978-1-57590-335-4

NACE International
The Worldwide Corrosion Authority
15835 Park Ten Place
Houston, TX 77084
nace.org

Contents

List of Symbols and Abbreviations

2LPE	Two-layer polyethylene
3LPE	Three-layer polyethylene
3LPO	Three-layer polyolefin
3LPP	Three-layer polypropylene
A-SCC	Axial stress corrosion cracking
AC	Alternating current
ACVG	Alternating current voltage gradient
ANSI	American National Standards Institute
API	American Petroleum Institute
ASTM	American Society for Testing and Materials
AWS	American Welding Society
AWWA	American Water Works Association
bpd	Barrels per day
BTD	Benzopherone tetracarboxylic dianhydride
C-SCC	Circumferential stress corrosion cracking
CEPA	Canadian Energy Pipeline Association
CFR	Code of Federal Regulations
CIS	Close interval survey
CP	Cathodic protection
CPE	Constant phase element
CSA	Canadian Standards Association
CSE	Copper sulfate electrode
CTE	Coal tar enamel
DC	Direct current
DCVG	Direct current voltage gradient
DFT	Fry-film thickness
dicy	Dicyandiamide

DIN	Deutsches Institut füer Normung
DOT	Department of Transportation
DSC	Differential scanning calorimeter
ECDA	External corrosion direct assessment
EIS	Electrochemical impedance spectroscopy
EMAT	Electromagnetic acoustic transducer
ESC	Environmental stress cracking
EVA	Ethylene-vinyl acetate
FBE	Fusion bonded epoxy
FMEA	Failure mode and effect analysis
FRA	Frequency response analyzer
FTIR	Fourier Transform Infrared
GPS	Global positioning system
HDPE	High-density polyethylene
HPCC	High performance composite coating
HVAC	High-voltage alternating current
HVDC	High-voltage direct current
ICCP	Impressed current cathodic protection
ILI	In-line inspection
ISO	International Organization for Standardization
LDPE	Low-density polyethylene
LEIS	Localized electrochemical impedance spectroscopy
LEL	Lower explosive limit
LLDPE	Linear low-density polyethylene
LPR	Linear polarization resistance
LNG	Liquefied natural gas
MAC	Maximum allowable concentration
MDPE	Medium-density polyethylene
MFL	Magnetic flux leakage
MIC	Microbiologically influenced corrosion
MLPP	Multi-layer polypropylene
MSDS	Material safety data sheets
NACE	National Association of Corrosion Engineers
NIOSH	National Institute for Occupational Safety and Health
NPS	Nominal pipe size
OSHA	Occupational Safety and Health Administration
PDL	Pipe diameter length
PE	Polyethylene
PIM	Pipeline integrity management
PP	Polypropylene
PPE	Personal protective equipment
PVDC	Polyvinylidene chloride
RH	Relative humidity
ROW	Right of way
SACP	Sacrificial anode cathodic protection
SCC	Stress corrosion cracking
SCE	Saturated calomel electrode

SHE	Standard hydrogen electrode
SKP	Scanning Kelvin probe
SRB	Sulfate reducing bacteria
SSPC	Society for Protective Coatings
TMA	Trimellitic anhydride
UHMWPE	Ultra high molecular weight polyethylene
UT	Ultrasound testing
UV	Ultraviolet
VEC	Viscous elastic coatings
WFT	Wet-film thickness
a	Constant
A	Area
A_d	Disbonded area of a coating
b	Constant
c	Concentration
C	Capacitance
C_c	Coating capacitance
C_{dl}	Double-layer capacitance
d	Distance
D_d	Delamination ratio
E	Potential
E_0	Amplitude of potential
E_{corr}	Corrosion potential
E_{CP}	CP potential
$E_{Fe2+/Fe}^0$	Standard equilibrium potential of iron
E_H	Potential for hydrogen evolution
E_{pit}	Pitting potential
E_{ref}	Potential of reference electrode
f	Frequency
f_b	Breakpoint frequency
f_c	Coating damage factor
f_{cf}	Final coating damage factor
$f_{c,i}$	Initial coating damage factor
f_{cm}	Mean coating damage factor
f_{min}	Minimum of frequency
F	Faraday's constant
I	Current
I_0	Amplitude of current
i	Current density
i_p	Passive current density
L	Thickness of a coating film
L_{in}	Inductance
p	Partial pressure
P	Permeability

Q	Amount
R	Ideal gas constant
R_{ct}	Charge-transfer resistance
R_e	Resistance
R_p	Polarization resistance
R_{pore}	Pore resistance
R_s	Solution resistance
S_w	Solubility
t	Time
t_f	CP design life
T	Temperature
T_g	Glass transition temperature
V	Volume
V_c	Surface potential
W	Tungsten
Y_0	Modulus
Z	Impedance
Z_0	Amplitude of impedance
Z_{CPE}	Impedance of a constant phase element
Z_{eq}	Equivalent impedance
Z_{ima}	Imaginary part of impedance
Z_{rea}	Real part of impedance

α	Exponent
$\hat{f}$	Measuring frequency at the peak of the semicircle in Nyquist diagram
φ	Phase angle
ϕ_{min}	Minimum of phase angle
ϕ_W	Work function
Φ_{probe}	work function of the Kelvin probe
$\Delta\Phi$	Contact potential
$\Delta\Phi^i_j$	Contact potential at individual interface between *i* and *j* materials
$\Delta\Phi_{domain}$	Donnan potential
$\Delta\psi_{probe\ coating}$	Kelvin potential measured on a coated metal specimen
ω	Radial frequency
ε	Dielectric constant
ε_0	Dielectric constant of free space
v	Volume fraction
ρ	Resistivity of coating
κ	Solution conductivity
$\chi_{coating}$	Surface potential of a coating
μ_e^{steel}	Chemical potential of a steel electrode

1 Introduction

1.1. Pipelines and Pipeline Integrity Management

Pipelines have effectively and efficiently transported large quantities of crude oil, natural gas, and diluted bitumen from production sites (usually remotely located) to refineries and markets. Compared to other transport modes such as rail, truck, and boat, pipelines are safer, more economic, and emit less carbon as they transport cargo across provinces, countries, and continents [Behar and Al-Azem, 2015]. With rapidly growing global demands for energy, oil and gas production has expanded substantially due to major technological advances. This expansion drives the increased need for new pipelines. For example, the U.S. is expected to achieve an average of 12.2 million barrels per day (bpd) with the production of oil, liquefied natural gas (LNG), and condensates, making it the world's largest producer of combined crude liquids [Cope, 2014]. In Canada, it is forecasted that by 2018, approximately 3.37 million bpd oil sands will be produced [Cope, 2015]. As a result, various ambitious plans have been proposed for new-build and expansions of pipelines to gather oil/gas products for delivery to markets.

Great effort has been made by multiple parties including pipeline operators, regulators, researchers, and society to keep pipelines away from risks of degradation and failure that could cause catastrophic consequences, such as energy loss, environmental and ecological damage, and even fatal accidents. Indirect negative impacts (e.g., public image, market share of pipeline companies, etc.) are difficult to estimate. Therefore, the management of pipeline system integrity and safety has been the fundamental and core business for all pipeline operating companies.

Pipeline integrity management (PIM) is the process to develop, implement, measure, and manage a pipeline's integrity through assessment, mitigation, and prevention of risks. The PIM ensures a safe, environmentally responsible, and reliable service [Nelson, 2002]. The importance of the PIM program is obvious. It can maintain the safe and reliable operation of pipelines for energy transport, improve pipeline system sustainability, and reduce operating risks by optimizing operational and capital expenditures, maximizing pipeline life cycle and reliability, and managing potential risks and threats. It also increases shareholder and public confidence in pipelines.

Generally, a PIM program consists of several interrelated modules (i.e., identification of potential risks for specific pipeline segments or the whole system, assessment of possible failure modes and associated consequences, implementation of preventive actions and mitigation measures, and recommendations for further program improvement). The design and implementation of the PIM program is highly specific and must consider actual conditions where a pipeline is operated. For example, the long-distance transmission of oil and gas through pipelines is usually subject to threats from external environments. As a result, attention should be paid to monitor, mitigate, and prevent external risks. For upstream-gathering pipelines, the carried fluid is usually highly corrosive and can also be erosive when a high content of solid sands is contained. Risks of pipeline failure are primarily internal. Thus, integrity management should focus on potential internal risks.

The PIM program usually includes five steps to maximize pipeline integrity and safety [Focke, 2015]. These include:

1. **Data gathering and alignment**. Pipeline operators collect all relevant data from various sources to the pipeline, including its design, construction, coating and welding, in-line inspection (ILI), cathodic protection (CP) monitoring, maintenance, repair, etc. The data identify existing critical features along the pipeline for scheduling rehabilitation measures. Moreover, data from any single inspection and monitoring cycle should be compiled and compared with data collected from previous inspections/monitoring of the same segment. Accurate data alignment is required for further improvement of the PIM program and pipeline integrity.
2. **Feature assessment**. After relevant data are collected and filed in a data processing system, it can be used to calculate technical parameters (for example, the maximum allowable operating pressure, the growth rate of the features, coating and CP performance, remaining service life of the pipe, integrity of welds, etc.). Data will also be analyzed for irregularities such as flaws, metal loss, cracks, etc. Established models and the comparison between historical data records allow for mechanic and quantitative analysis. Each identified feature is to be assessed separately.
3. **Condition analysis**. In condition analysis, inspection data and calculated parameters are used to generate a ranking or an index number that determines a pipeline's fitness-for-service. The risk of failure can be estimated for individual inspection features, for pipe segments, or for a whole pipeline.
4. **Risk assessment**. Risk assessments consider the probability of failure occurring on pipeline segments and the potential consequences to public safety, the environment, and operators' financial stability.
5. **Integrity planning**. For identified features and potential risks, integrity planning is conducted by relevant parties to address pipeline issues. Planning action measures is included in the work management systems.

1.2. Coatings for Pipeline Corrosion Prevention

Corrosion is one of the key mechanisms affecting the durability and integrity of pipelines. Coatings in conjunction with CP provide the primary means to protect a pipeline from corrosion attack, mechanical damage, and geotechnical threats, and to maintain a pipeline's integrity in soil or water environments. In particular, the coating forms the first line of defense against a pipeline's external corrosion. However, a coating can degrade or fail at various stages of pipeline design, construction, and operation. Stages include coating manufacturing, application on pipes either in the plant or in the field, transportation, installation, and operation of the coated pipe. Moreover, the pipeline infrastructure around the world is aging. Statistics show [Hopkins, 2007] that over 50% of the oil and gas pipeline systems in the U.S. are over 40 years old, and 20% of Russia's oil and gas pipelines are nearing the end of their design life. Aged pipeline assets, including the coatings, become important challenges to the integrity of pipeline systems.

The principle of the strategy to combine coating with CP in PIM is that the coating, if it is intact and adheres well to the pipe's steel substrate, effectively separates the pipe from the environment, and at the same time, reduces the CP current demand. Where coating has failed, the CP acts as a backup to protect the pipeline from corrosion. However, when both coating and CP fail, the pipeline becomes susceptible to external corrosion and/or stress corrosion cracking (SCC). Industrial experiences have shown that coating failure is always the prerequisite for corrosion and SCC to occur on pipelines [National Energy Board, 1996; Cheng, 2013]. Due to its essential role in pipeline integrity maintenance, the coating has been integral to the PIM program and should be considered as a part of the whole pipeline system.

In the PIM program's five-step process as described earlier, coatings are involved in at least three steps (i.e., data gathering and alignment, condition analysis, and integrity planning). All data about the coating selected and applied to the pipeline (including its type and manufacturing, the plant-applied procedure, the field-applied coating and its compatibility with the main line coating, history of uses, performance status, periodic inspection records, etc.) should be collected and integrated into the PIM program. The data, especially the coating performance inspection results, will be analyzed along with other inspection data to evaluate the performance and status of the coating and the pipeline. Moreover, the compatibility of the coating with CP will be evaluated to determine the CP effectiveness once the coating has failed, such as when it disbonds from the pipe steel. Analysis results and the coating performance evaluation will guide the actions and rehabilitation plans required to improve pipeline integrity.

Generally, factors to be considered during the selection and design of pipeline coatings include but are not limited to:

- mechanical properties of the coating
- chemical and electrochemical properties of the coating
- susceptibility to coating damage with pipe handling during installation and repair
- soil chemistry
- compatibility for in-situ joint coating

- coating compatibility with CP
- estimated service life of the coating

All of these can affect pipeline integrity and thus the safety of the pipeline system. In addition to their corrosion resistance, selected coatings for pipeline use must be resistant to mechanical damages resulted from pipe handling, trench backfill, soil conditions, and the suitability of field joint coatings. The coating must serve as an effective barrier that separates the pipeline steel from the environment, providing long-term pipeline protection. It must remain intact and adhered, assuring both corrosion resistance and mechanical strength.

In summary, an ideal pipeline coating should be worker-safe, environmentally friendly, durable, and able to seal all substrate metal surfaces from the service environment. It must also be resistant to environmental, mechanical, and chemical damage during application, handling, burial, and service. It should be applied efficiently and effectively under the restricted environmental and work conditions in the field. Finally, it should come at a reasonable cost, even though cost should not be the main decision point in coating selection.

A wide variety of coatings have been used for corrosion protection and integrity maintenance for oil and gas pipelines over the last several decades [Niu and Cheng, 2008]. These include coal tar, asphalt, polyethylene (PE) coatings, fusion-bonded epoxy (FBE) or dual layer FBE coatings, three or multi-layer polyolefin (PE or polypropylene) coatings, composite coatings, etc. Although most of these coatings have successfully maintained pipeline integrity, challenges remain for the industry with various exceptional applications as well as oil/gas production activities conducted in increasingly remote, geographically difficult areas. These include extremely cold weather, unstable geotechnical conditions (such as slopes, earthquake zones, permafrost or semi-permafrost, etc.), microbial activity, and water and gas permeability over the long term. The industry has long pursued novel and effective pipeline coating technologies to meet these integrity-related challenges.

1.3. Contents of the Book

The evolution and development of pipeline coating technology can be traced to the 1940s and continues to be of global interest. Currently, design, selection, application, uses, and management of coatings has been integrated with pipeline systems' PIM programs. Our understanding of pipeline coatings has evolved to a stage that delivers a comprehensive review describing the scientific, technical, and practical aspects of pipeline coatings. All of these facilitated the development of this book.

The book begins with a review of coating fundamentals in Chapter Two, where the evolution of coating technology and the principles for coating formulation are introduced. Guidelines for coating design, selection, and application are briefly presented. The structure of a coating system and the high-performance coating's essential properties and characteristics are covered in detail. Some standard testing methods for determining and evaluating coating properties are included.

Coatings used in the oil/gas pipeline industry are covered in Chapter Three. Generally, pipeline coatings are divided into two categories (plant-applied and field-applied coatings). The chapter describes primary coatings in both categories such as coal tar, asphalt, PE, liquid epoxy, FBE, and high-performance composite coating (HPCC), as well as field-applied liquid coatings (i.e., tape coatings, shrink sleeve, wax, mastics and many others) in terms of their structures, properties, products, and applications.

Coating failures encountered on pipelines in the field are reviewed in Chapter Four, which includes an analysis of its effect on pipeline integrity. Both permeable coatings and impermeable coatings receive particular attention, and their interactions with CP are discussed. The shielding effect of coating failures under a variety of scenarios is included to provide an understanding of this industry-important problem. The tests and results described in this chapter come from the authors' research activities. This first-hand information provides recommendations to the industry for avoiding incompatibility between pipeline coating candidates and CP.

SCC has been a primary mechanism resulting in pipeline failure [National Energy Board, 1996]. It has been acknowledged [Cheng, 2013] that SCC occurrence is subject to coating failures. Chapter Five focuses on mechanistic aspects of the essential role of coating failures in pipeline SCC, including its initiation and propagation. Both near-neutral pH and high-pH SCC on pipelines are introduced, and correlations between the type and properties of coatings and their failure mechanisms are established. Discussions detail the development of solution chemistry and electrochemistry under disbonded coating to support SCC. Again, the majority of the results discussed in this chapter come from the authors' research experiences. After following this content, readers may connect coating performance with the potential occurrence of pipeline SCC.

Techniques for characterizing coating properties and testing coating performance in the field and research laboratory are covered in Chapters Six and Seven, respectively. The discussion provides insights essential to a complete testing and evaluation program for pipeline coating candidates, and for predicting long-term coating performance. Depending on an individual technique's capability and actual coating property needs, one can choose a testing method from convenient, simple inspection tools to complex, research-oriented equipment.

Various coating application techniques are introduced in Chapter Eight, which covers almost all important issues required for understanding what is necessary when a coating is applied to metal substrate, including pipelines. The content is based on realistic experiences.

Finally, industrial experience with inspection and management of pipeline coatings is included in Chapter Nine. Inspections have been integral to the PIM program and ensure the integrity and safety of pipeline systems.

References

Behar, J., Al-Azem, S. (2015) Pipelines, oil boom and railroads, *World Pipelines* 15 (4), 18-28.

Cheng, Y.F. (2013) *Stress Corrosion Cracking of Pipelines*, John Wiley Publishing, Hoboken, NJ, USA.

Cope, G. (2014) Growing pains, *World Pipelines* 14 (1), 12-16.

Cope, G. (2015) A cloud of uncertainty in Canada, *World Pipelines* 15 (6), 14-18.

Focke, J. (2015) The future of pipeline integrity management, *Pipelines Inter.* Issue 3, 28-29.

Hopkins, P. (2007) Pipelines: past, present, and future, the 5th Asian Pacific IIW International Congress, Sydney, Australia, Mar. 7-9.

National Energy Board (1996) *Stress Corrosion Cracking on Canadian Oil and Gas Pipelines*, MH-2-95, Calgary, Canada.

Nelson, B.R. (2002) Pipeline integrity: program development, risk assessment and data management, the 11th Annual GIS for Oil & Gas Conference, Houston, USA.

Niu, L., Cheng, Y.F. (2008) Development of innovative coating technology for pipeline operation crossing the permafrost terrain, *Constr. Build. Mater.* 22, 417-422.

Coating Fundamentals

2.1. Evolution of Coating Technology

Before moving into more technical content, it is worth understanding the difference between two concepts that are often considered as one but do have different meanings, i.e., paint and coating. A paint is defined as any liquid material containing only drying oil or in combination with natural resins and pigments that combine with oxygen in the air to form a solid, continuous film over the substrate, providing a weather-resistant, decorative surface [Munger and Vincent, 1999]. Generally, paints do not provide permanent protection against corrosion for structures underneath. A coating is a material composed of synthetic resins or inorganic silicate polymers that provide a continuous film capable of resisting industrial, marine, and other corrosive environments [Munger and Vincent, 1999]. Compared to paints, coatings are processed in more complicated mechanisms and offer outstanding adhesion, mechanical strength, and resistance to water, chemicals, humidity, and weather, as well as other properties.

In prehistoric times, paints were primarily used for art. It was believed [PSG Web Source] that paints made their earliest appearance approximately 30,000 years ago when cave dwellers used crude paints on walls to depict their lives. The paints were made of natural substances, such as earth pigments, iron oxides, berry juice, lard, blood, and milkweed sap.

In ancient times, long before its application for protection to the substrate structures, paint was mainly used for decorative and identification purposes. In Egypt, the first synthetic pigment known as Egyptian Blue (made by heating lime, soda ash, sand, and copper oxide and ground into a fine powder) was developed in 8000-6000 BC [World Book Encyclopedia, 1978]. The Chinese, Koreans, and Japanese began using lacquer to decorate their buildings, instruments, and weapons in 6000 BC [Encyclopedia Britannica, 1974]. In the Roman Era, the Romans learned about making paint from the Egyptians, and developed some lead-based artificial colors. In 600 BC to AD 400, the Egyptians,

Chinese, Hebrews, Greeks, and Romans used oils as varnishes, pigments such as yellow and red ochres, chalk, and arsenic sulfide. They also used binders such as gum arabic, lime, egg albumen, and beeswax in the paints. Early Native Americans used a variety of organic materials for paint as well [World Book Encyclopedia, 1978].

In the 1500s, artists began to add drying oils to paints to hasten evaporation. They adopted a new solvent (linseed oil), which was the most commonly used solvent until synthetic solvents replaced it during the 20th century. In 1700, Thomas Child established the first recorded paint mill in Boston, USA [PSG Web Source]. The first paint patent was issued for a product that can improve whitewash, a water-slaked lime.

In the 1800s, early coating technologies started to emerge. Some important events include [PSG Web Source]:

- In 1856, Henry Perkins discovered the first real synthetic dye, pioneering the manufacturing of dyes synthetically and economically.
- In 1865, Flinn patented a water-based paint containing zinc oxide, potassium hydroxide, resin, milk, and linseed oil.
- In 1880, the Sherwin-Williams Company improved the suspension of fine particles of linseed oil, making its paints the best of all paints available at that time.

In the early 1900s, protective coatings were developed to meet the immediate needs of railroad construction [Munger and Vincent, 1999]. Corrosion of the riveted steel bridges that were vital to the railway threatened the safety and reliability of the whole system. This drove the development and uses of a new corrosion-resistant coating system for the first time in history: the one consisting of a red lead-linseed oil primer applied in one or more coats, followed by a linseed oil-graphite topcoat. This truly protective coating performed very well under the railway's service condition in many environments, except in highly corrosive marine and industrial environments.

The first completely synthetic resin was made from phenol-formaldehyde in the early 1900s [Knop and Pilato, 1986; Bedard and Riedl, 1990]. The resin's increased flexibility decreased drying time, making it more weather- and water-resistant. The continued development of solvents and plasticizers, together with the synthetic resin, made high-performance protective coatings possible.

A major breakthrough in the development of corrosion-resistant coatings was the production of chlorinated rubber-vinyl copolymer combinations in the early 1940s [Munger and Vincent, 1999]. This development overcame the problem that when used alone, chlorinated rubbers were extremely brittle and hard. Adding alkyd resins to the rubber increased chlorinated rubber coating's plasticity and improved its color and gloss retention properties. As a result, the first chemical-resistant protective coating combined chlorinated rubber (as primer) with a vinyl copolymer (as body or top coat). This coating system was applied in the chemical, sewer, and marine industries to protect the substrate structure.

The first inorganic zinc coating was developed in Australia by Victor Nightingall, Australia's Edison, in the 1930s [Francis, 2013]. Inorganic zinc coatings were quite different from any other coatings.

The liquid part, or binder, consisted of a silicate (e.g., sodium silicate) in early formulations, and potassium, lithium, and ethyl silicate more recently. A large amount of zinc was added to the binder during stirring. The coating was then applied to a clean, abrasively blasted steel surface. The coating dried and hardened quickly, but did not initially resist water permeation. The coating continued to harden over the next six to twelve months. A number of chemical reactions occurred at different rates during this time period. At almost the same time, organic zinc-rich coatings were developed in Europe [Schweitzer, 2005]. The advantage of organic zinc-rich coatings is the galvanic protection offered by the coating's zinc component, due to zinc's high electrochemical activity compared to many other metals. Organic coatings are more tolerant of surface preparation than the inorganic zinc coatings. However, both inorganic and organic zinc coatings have contributed to the use of zinc primers for almost all high-performance protective coatings.

Epoxy resins were developed shortly after World War II and had a major impact on coating technology innovation. Epoxy resins have good adhesion and corrosion resistance. Moreover, the resins can be applied to structural surfaces easily. Soon after epoxies were developed, polyamide epoxy coatings were found to have increased adhesion, flexibility, and water and chalk resistance. Polyurethane coatings were also developed during this period, but were considered inferior to epoxy with regard to water resistance. However, epoxy resins suffered from poor resistance to ultraviolet (UV) attack, resulting in chalking and loss of gloss and color. It was not until two-component aliphatic polyurethane topcoats were created in the 1960s, when the color and gloss retention of epoxy coatings was improved, the top coat's chemical resistance was increased [Munger and Vincent, 1999]. Currently, there are many epoxy coatings with numerous curing agents available for structural protection.

During the past several decades, protective coatings have undergone significant technological development. Coatings have evolved from coal tar and asphalt to advanced single- and multiple-layered or composite coatings [Niu and Cheng, 2008]. In addition to a wide variety of plant coatings, advances in field application (due to improved manufacturing processing and advancement of material sciences) have allowed joint coatings in the field to be more robust and easier to apply. Another breakthrough has been with the development of next-generation, heat-shrinkable sleeves for three-layer polyethylene (3LPE) and new polypropylene (PP) heat-shrinkable sleeves for multi-layer polypropylene (MLPP). These sleeves allow effective applications at temperatures lower than those in manufacturing, but still provide consistent mechanical performance, chemical resistance, and good thermal performance [Buchanan, 2003].

With the advancement of nanotechnology, various nano-coatings have recently been developed. Particularly, "smart coatings" refer to coating systems with corrosion-sensing and self-healing properties, providing not only an environmental barrier, but also the "smart" release of corrosion inhibitors, which are preloaded into the coating as demanded by coating damage/degradation and the presence of a corrosive environment on metal [Kumar and Stephenson, 2004; Bohannon, 2005]. Application of smart coatings for corrosion protection possesses a number of advantages, including improved corrosion inhibition suitable for specific environments, intelligent coating systems capable of sensing corrosion onset, environmental compliance, and cost effectiveness. As an innovative technology, smart coatings have been developed for protection of structures primarily made of light metals such as aluminum and magnesium alloys [Lamaka et al., 2007]. While very limited work has been conducted on steels (including carbon steels) [Kumar et al., 2006], no smart coating technology has been developed for pipeline applications.

2.2. Principles of Coating Formation

Understanding the principles of coating formation is critical to the use of coatings for pipeline integrity management. Coating formation is a complex process that involves multiple steps of physical and chemical reactions. Of all sub-reactions, the most critical one governs a coating's conversion from liquid to solid. This is discussed in detail in the following sections.

The type of coating film formulations depends on the relationship between the resin's molecular weight and its dissolvability. Generally, the former is inversely proportional to the latter. Resins with increased molecular weights or sizes are associated with decreased solvability and increased viscosity. As a result, coating films formed by big molecules are usually based on evaporation rather than a series of chemical reactions. It is difficult to apply a coating to the substrate surface. Smaller and less complex resins with smaller molecular weights are more soluble in solvents. This allows polymerization and/or oxidation reactions to occur, and sometimes through the addition of catalysts to form complex cross-linked resins that form coating films. These complex, cross-linked coatings are usually more resistant to corrosion.

2.2.1. Coating Film Formation by Solvent Evaporation

Resins that dry by solvent evaporation are all thermoplastic film formers, such as vinyl resins [Munger and Vincent, 1999]. Although solvent evaporation seems to be a simple principle for coating formation, the processes involved are complicated. The film-forming process starts only when solvent evaporation reaches an advanced stage, so that resin molecules are close to each other and allow the chemical attraction to become substantial. The formed films have different properties, depending on molecular structure. For example, a homogeneous, dense structure is generated by a solvent that facilitates maximum polymer dispersion and mobility during film formation. If not, polymer accumulation may occur, resulting in a poor coating property.

The quality and quantity of solvents is critical to homogenous coating films. If the solvent evaporates too fast, the resin tends to dry before it applies to the structure. If the solvent evaporates too slowly, a slow film formation is caused and the coating film can remain tacky, making it less resistant to chemical and corrosion attack. It is common to use a combination of solvents having various evaporation rates to form the film. The different evaporation rates create conditions for resin molecules to orient themselves properly and form a smooth, continuous film.

2.2.2. Film Formation by Oxidation

A coating film formed by oxidation applies primarily to dry oils, which are placed on the surface of substrates in a thin film for a certain time period until that they have reacted with oxygen in the air to become dry and hard. Initially, the oils are resistant to atmospheric conditions. In time, they become hard, and eventually crack and chip away from the substrate surface.

Long-chain unsaturated oil molecules react with oxygen to isomerize, polymerize, and cleave the carbon-carbon chain, as well as form oxide products. Oxygen uptake is an important factor during the process, which can be divided into four separate steps [Munger and Vincent, 1999]:

1. There is no visible change in physical and chemical properties of the oil resins during the induction period, but antioxidants in the film are being destroyed.
2. Oxygen uptake is appreciable, and hydroperoxides and conjugation form.
3. Decomposition of hydroperoxides forms free radicals, and the oxidation reaction is autocatalytic.
4. Polymerization and cleavage reactions occur. Complex, high-weight cross-linked polymer films are formed. At the same time, low molecular weight products are also formed. The oxygen absorption rate reaches a maximum at this point. After that oxygen continues to be absorbed, but at a slow rate [Craver and Tess, 1975].

2.2.3. Film Formation by Polymerization

Polymerization usually occurs between a monomer and one or more polymers of different types to produce a resin film that is cross-linked with a rigid, three-dimensional molecular structure. The coating formed is thermoset (i.e., the coating is insoluble in its own solvents and is not appreciably softened by heating). Coatings formed by this principle are much more resistant to corrosion. Moreover, they usually possess a high hardness, and are both chemically and water resistant. As the polymerization process must be controlled under strict manufacturing conditions, it is not practical to form polymer coatings on-site.

Frequently, polymerization processes require certain catalysts to form copolymer resins from unsaturated polymer molecules. The catalyzed polymerization can generate thick films, with good water and chemical resistance.

2.3. Structure of a Coating System

In general, a coating system consists of multiple layers: primarily a coat of primer, an intermediate coat (or body coat), and a top coat. Figure 2.1 shows different layers of a coating system [Munger and Vincent, 1999]. Each layer is functional for certain purposes to ensure that the whole system maximizes its protection from corrosion attack to the substrate structure.

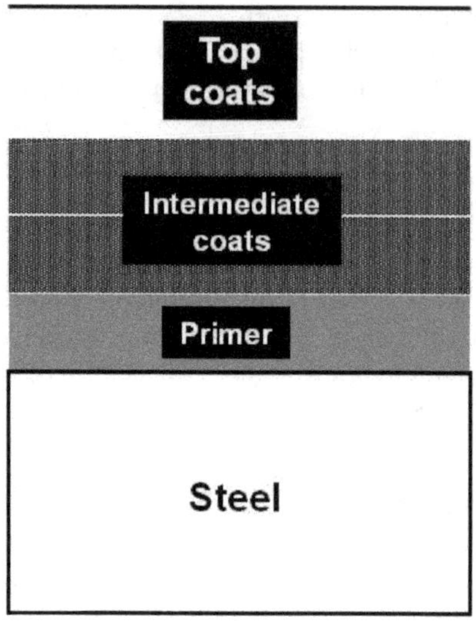

Figure 2.1. Three different layers in a coating system [Munger and Vincent, 1999].

2.3.1. Primer

Primer is one of the most critical parts of a coating system and is universal for all corrosion-resistant coatings. Primer is the coating layer immediately next to the substrate, and has essential functions, including:

- Adhesion: Primer provides a strong bond between the coating and substrate. In fact, a primer is the base to which the rest of the coating system is applied. Therefore, it must have a strong adhesion to the substrate. Adhesion is often regarded as the most important property the primer possesses.
- Cohesion: Primer must maintain a strong bond between coating layers and provide internal strength. It should be compatible with the body coat and provide a strong adhesion to the intermediate coat.
- Corrosion resistance: In reality, when primers are applied, the substrate usually stands for many days or even months before a coating is applied. The gap between when the primer is applied and when the topcoat is applied requires the primer to have a good corrosion resistance to the environment.
- Resistance to water permeation: The primer is the last line of defense to penetration of water or moisture through the coating system. It must be sufficiently resistant to moisture permeation during service.
- Carriage of corrosion inhibitors: Primers can be used alone as a single-layer coating. For effective corrosion protection, the primer can be loaded with pigments, which can inhibit corrosion of substrate metal (or passivate the metal) once corrosion does occur. A cathodically active primer must react with moisture and electrolytes from the environment to cathodically protect the metal substrate [Munger and Vincent, 1999].
- Distention: The primer should possess appropriate flexibility to maintain integrity during mechanical bending of the substrate structure, or when the temperature changes in seasonal cycles.

2.3.2. Intermediate Coat (or Body Coat)

An intermediate coat (or body coat) in the coating system is usually used for specific purposes. It is not universal, and cannot be used as a finishing coat. The main functions of the body coat include:

- Thickness of the total coating system: The main function of the body coat is to increase a coating's thickness. Generally, increasing the coating thickness can improve many other properties, such as increased chemical resistance, improved resistance to moisture-vapor transfer, increased electrical resistance, and enhanced impact and abrasion resistance. Moreover, a coating with just average properties can make up additional properties with an increased thickness.
- Chemical resistance: The body coat can increase the coating's chemical resistance by acting as a barrier to permeating chemicals in corrosive environments.
- Increased resistance to electricity and improved impact and abrasion resistance: These properties depend on an increased coating thickness.
- Reduction of the rate of water or moisture permeating the coating.
- Strong bonding to primer and top coat: An intermediate coat usually has a high pigment content so that it is a flat coat with a good physical adhesion [Munger and Vincent, 1999].

2.3.3. Top Coat

Top coat is the coating system layer in direct contact with service environments. It is denser and thinner, and contains a lower volume of pigments compared to the body coat. The important functions of the top coat include:

- Provides a blocking barrier to the environment. The top coat is the first line of defense against an environmental attack to the substrate. It must be a resistant seal for the coating system.
- Improved resistance to chemical, moisture and gas permeation.
- Provides a tough and wear-resistant surface.
- Provides color, gloss, and texture, with appearance features.
- Provide protection from ultra-violent ray deterioration (in above-ground installations).

For structures used in marine environments, the top coat may also provide resistance to marine fouling. In some cases, the top coat is applied for appearance only.

2.3.4. Mixed Coating Systems

A coating system with multiple coat layers can be effective to protect against corrosion attack of the substrate structure. However, the coating system needs not include all three parts (as described above). Even a single coat layer can be highly functional for corrosion protection. Coating systems may consist of any number of coats and combinations of materials. Additional layers and components may be designed for specific purposes.

Use of a mixed-coating system is usually not recommended, as it frequently causes problems. For example, with the application of vinyl or epoxy coatings over an alkyd or shop primer, solvents from the top coat penetrate and break up the alkyd so coating integrity is lost. Thus, mixed-coating systems should be investigated completely prior to use.

2.4. Coating Components

As stated, coatings may contain several layers and a number of components that are manipulated to alter coating properties to meet service requirements for a specific project. Moreover, individual layers can include various components to help form desired properties. Some primary components in a coating system are introduced in the following discussion.

2.4.1. Binders

Binders are materials that convert a coating into a dense, solid, and adherent membrane. Binder's ability is related to its molecular size and complexity. Generally, binders with a high molecular weight form coating films by evaporation of solvents only, and the resin or the binder molecule is in its completed form prior to application for coating formation. Other binders must chemically react in place.

A number of different binders are available for coating formulation including [Munger and Vincent, 1999].

Oxygen-reactive binders. Generally, oxygen reactive binders are low molecular weight resins, which can produce coatings through an intermolecular reaction with oxygen, usually through the catalytic effects of metallic salts such as cobalt and lead. Some typical coatings formed by oxygen reactive binders include: (1) *alkyds*, natural drying oils chemically reacted into a synthetic resin; (2) *epoxy esters*, epoxy resins combined chemically with drying oils; (3) *urethane alkyds*, epoxy resins chemically combined with drying oils as part of the molecule that is further reacted with isocyanates; (4) *silicone alkyds*, the weather-resistant binder formed by the combination of alkyd resins with silicone molecules.

Lacquers. Lacquers are coatings that form when solvents evaporate from a liquid into a solid film. Lacquers generally have a lower volume of solids than materials formed from lower molecular weight resins. Asphalts and coal tars are often combined with solvents to form lacquer-type films, providing good chemical- and corrosion-resistant films.

Heat-conversion binders. The heat-conversion binders are generally used as basic coating materials and are not combined with any other resins. Powder coatings are made of heat-conversion binders, which can be high molecular weight thermoplastic resins, such as epoxies. Resins are usually converted to a fine powder and applied to a heated substrate.

Co-reactive binders. Co-reactive binders are formed from two low molecular weight resins, which, when combined just before application, co-react with each other and form a solid film after being applied to the structure. Two important coatings made of co-reactive binders are epoxies and polyurethanes. Epoxy binders consist of relatively low molecular weight resins, which react with ammonia-type compounds such as amines and form the solid binder. Polyurethanes are co-reactive binders with urethane prepolymers that react with resins or chemicals containing amines or alcohols to form the finished coating.

Condensation binders. The binders are based primarily on resins that interact to form cross-linked polymers when subject to high temperatures. Condensation is the release of water during the polymerization process. The condensed materials are strongly cross-linked and are very chemical-resistant.

Inorganic binders. Inorganic binders are primarily inorganic silicates that are dissolved in either water or solvents, and react with moisture or carbon dioxide in the air to form an inorganic film. The type of inorganic binders depends on the form of the silicate during its curing period, such as post-cured inorganic silicates, self-curing water-based silicates, and self-curing solvent-based silicates.

To select appropriate binders during the design of a coating system, it is important to understand the advantages and disadvantage of various types of binders. Table 2-1 summarizes relevant information of some typical binders used in coating processing.

Table 2-1. Advantages and Disadvantage of Some Typical Binder Types [Zaki, 2006].

Binders	Advantages	Disadvantages
Drying oils	Economical and easy to apply	Poor corrosion resistance
Alkyds	Good weather resistance	Poor corrosion resistance
Epoxy esters	Highly corrosion resistant	Expensive and poor weather resistance
Chlorinated rubber	Resistant to oils, acids and alkalies	Limited to applications <80°C
Bituminous materials	Economic, good water and rust resistance, strongly adhesive	Limited to applications <65°C
Polyurethane	High corrosion resistance	Can discolor
Silicates	Good weather and chemical resistance	

2.4.2. Solvents

Solvents are low molecular weight organic compounds that reduce the viscosity of other solids or fluids. In coatings, solvents dissolve the binder to generate desired properties. Although solvents do not remain in the coating after it forms, they affect the coating by creating porosity, discoloration, floating of pigment, fisheyes; by reducing coating strength; and by degrading the adhesion to substrate. A proper use of solvents creates a smooth, clear resin film with a good gloss. Moreover, a coating made with properly used solvents will have the inherent strength and other favorable properties of the basic resin.

There are few coatings that use only a single solvent. The majority of coatings are made with a combination of solvents. A combination of solvents usually provides better coating films. Each type of resin has a specific combination of solvents for generating the best coating. There is no universal solvent applicable to all coatings. Solvents are actually specific to binders.

Based on their solubility to binders, solvents can be divided into three types: active solvent, latent solvent, and diluent.

Active solvents. Active solvents are able to dissolve completely a resin to form a homogeneous solution. Active solvents are used for particular binders only. For example, ketone is an active solvent and completely dissolves vinyl resins; the aromatic hydrocarbon solvent is used for chlorinated rubber.

Latent solvents. Latent solvents may only swell a binder at room temperature, but form a solution at high temperatures. When the solution is cooled down, a gel is formed. Latent solvents can be used with active solvents to adjust the solvents' evaporation rate, improving the film properties.

Diluents: Diluents are not true solvents for resins. When combined with active solvents, they can dilute the solution. Diluents improve film properties, such as flexibility, to provide a smoother and stronger film and to reduce cost.

Solvent compatibility must be considered in the use of multiple solvents. A solvent combination is often used to improve various resins' compatibility and result in a good film. Due to different evaporation rates, one solvent can dissolve into the resin while others evaporate. This causes solvent retention in the coating, which reduces adhesion and the coating's water and chemical resistances. It can also result in blistering of the film.

Two basic categories of solvents in common use are hydrocarbons and chemical solvents. Typically, chemical solvents have a higher boiling point than that of hydrocarbon-based solvents. This reduces the solvent's evaporation rate, allowing it to spread over a surface more evenly.

Recent developments in coating formulas limit or eliminate solvents. Solid coatings can be less hazardous to health, have lowered environmental impact, and reduce the amount of vapors and odors that workers are exposed to. Mechanical and physical properties, such as permeability, abrasion, and impact and wear resistance can also be improved when solvents are eliminated from the coating system. A typical example is the group of solid polyurethane coatings [Guan, 2001].

2.4.3. Pigments

Pigments are added to coating systems for specific functions. Different pigments may be used in one coating, playing different roles that affect that coating's properties. For example, pigments can provide pleasing color and decorative characteristics and hide the substrate. They can protect resin binders by absorbing and reflecting solar radiation. Inhibitive pigments, such as borates, phosphates, and molybdates are principally used in primer to passivate the metal substrate for corrosion protection. Reinforcement pigments are finely divided fibrous and plate-like particles. They increase the hardness, toughness, and tensile strength of the coating film. Pigments can also increase coating adhesion, compared to primers without pigments.

The volume of pigments added to resins is specific to the resin-pigment combination. Ideally, pigment particles should be distributed uniformly in the binder resin matrix. If the pigment amount exceeds the critical concentration, the coating film becomes porous and insufficiently strong.

Pigments are categorized into coloring, reinforcing, inhibiting, or metallic pigments. As mentioned earlier, the coloring pigment offers pleasing color and decorative characteristics. Moreover, it protects resinous binders from the sun's UV-ray penetration. Reinforcing pigments improve the coating film's ductility to make it tougher and less likely to crack after service under adverse weather conditions. Reinforcing pigments also increase a film's hardness, strength, and chemical resistance. Inhibiting pigments primarily react with the substrate and generate a passivated surface for corrosion inhibition, and can themselves function as corrosion inhibitors. Metallic pigments generally appear as flakes or flat platelets, and improve a coating's adhesion and reinforce the binder.

2.5. Coating Properties and Characteristics

Inadequate coating performance is a major contributor to increased corrosion and SCC susceptibility of underground pipelines. Most external corrosion on pipelines is caused by disbonded coatings that shield CP, not lack of CP [Norsworthy - 2009]. A coating's key function is to separate a protected substrate from the service environment and prevent its contact with active, corrosive industrial fumes, liquids, solids, or gases. Coatings possess a wide variety of physical, chemical, and mechanical properties, which affect a coating's function and service life. It is essential to have a complete understanding of a coating's properties before using it. Desired pipeline coating characteristics include:

- Effective electrical insulation
- Effective moisture barrier
- Good adhesion to pipe steel
- Applicable by a method that will not adversely affect pipe properties
- Applicable with a minimum of defects
- Ability to resist generation and development of holidays over time
- Ability to resist damage and deterioration during handling, storage, and installation
- Ability to maintain substantially constant resistivity over time
- Resistance to disbonding
- Resistance to chemical degradation
- Ease of repair
- Retention of physical characteristics
- Nontoxic to environment
- Be compatible with CP and resist cathodic disbondment
- Resistance to soil stress

The following information describes some essential properties and characteristics that a coating possesses to achieve sustainable, high-performance use in-service.

2.5.1. Water Resistance

Water resistance is probably the most important coating characteristic, since most other properties are related to it. Generally, small molecules (for example, water molecules) can penetrate organic materials. Water passes through intermolecular spaces in an organic material, either remaining in an absorbed state or passing through the material [Munger and Vincent, 1999]. Eventually, moisture comes to an equilibrium, with the amount of water that enters the coating identical to the amount of water that evaporates from the coating surface. Thus, the water content inside the coating tends to achieve a constant value that depends on temperature and moisture-vapor pressure. The majority of coatings (including pipeline coatings) are made of organic materials and should have the highest possible water and moisture resistance to maintain their properties and be effective over a long period of time.

Frequently, coated structures are used in environments containing various types of chemical ions, molecules, gases, etc. Water permeation can also affect the permeability of other species if their sizes are sufficiently small. For example, water vapor can aid small molecules (such as carbon dioxide, ammonia, etc.) during their penetration into organic coatings [Munger and Vincent, 1999]. The permeation of these species can degrade coating properties remarkably, especially the coating's adhesion and resistance.

Some important concepts relevant to the water resistance of coatings are described as follows.

Water absorption. Water absorption refers to the amount of water that is absorbed and retained within the coating's intermolecular spaces. Under certain temperature-pressure pairs, the water content in a coating is in equilibrium with the atmosphere, with the coating desorbing water under dry conditions and absorbing water in high humidity or when immersed. All coatings have their own levels of water absorption. When a coating is strongly adhered to substrate metal and there is no disbondment between the coating and the substrate metal, the moisture absorbed into the coating would not cause corrosion of the metal. At given moisture-vapor pressures, the number of absorbed water molecules in the coating is in equilibrium with that of water molecules desorbing out of the coating. Thus, the content of water molecules in the coating remains constant. Generally, the best corrosion-resistant coating has the lowest water absorption.

Moisture vapor transfer rate. The moisture vapor transfer rate is the rate at which moisture vapor transfers through a coating when there is a difference in moisture-vapor pressures on one side of the coating compared to the other side. Each coating has its own characteristic moisture-vapor transfer rate. In general, the lower the moisture-vapor transfer rate, the better the protection the coating can provide. The transfer of water or moisture through a resin film is through intermolecular space, rather than through physical imperfections such as pinholes, voids, etc. The latter assembles the situation that water or moisture penetrates through the opening with direct access to the substrate metal and either causes corrosion, degrades adhesion of the coating, or both.

Blistering. Moisture transfer through a coating depends on the difference of pressures between the coating's two sides. If the coating is tightly adhered to the substrate, there is no difference in pressures between both sides, and the coating is in equilibrium with the moisture in the air. However, if the coating is disbonded from the substrate, a gap is generated under the coating. Moisture vapor tends to transfer into the gap and condense locally. At elevated temperatures, the moisture vapor in the gap results in sufficient pressure to create a blister. The blistered area can expand with the penetration of moisture vapor into the disbonded gap.

Thermal gradient across coating. When the metal substrate is at a lower temperature than the moisture vapor or water existing on the exterior of the coating, the warmer moisture vapor can penetrate through the coating and tend to condense on the cooler steel underneath the coating. As a result, a water-filled blister is created.

Osmosis. Osmosis refers to the passage of water through a coating film from a less-concentrated solution to a greater-concentration solution. Since all organic materials are somewhat permeable by water or moisture, this mechanism applies to all organic coatings, especially when they are subject to immersion in aqueous solutions, condensation, or even high humidity. The surface condition of the substrate metal underneath a coating affects osmosis. Generally, moisture can be pulled through the coating towards an area of contamination. Thus, a clean substrate surface can reduce osmosis when a coating is immersed or subject to high humidity.

Electroendosmosis. Electroendosmosis is defined as the forcing of water through a coating film by an electric potential with the same charge as the coating. Coatings are generally negatively charged. At coating breaks or defects, excessive negative electrons accumulate, making the metal a negative surface. Water tends to permeate through the coating towards the metal, i.e., cathode.

This mechanism amplifies its effect on metal structures that are under CP, such as buried pipelines. The cathodically polarized pipeline steel is negatively charged for corrosion protection. However, the steel cathode can force water that exists in the soil environment to penetrate through the coating towards the steel surface, resulting in either coating degradation or corrosion of the steel under the coating as the coating is disbonded from the steel.

With certain coatings this is an advantage. FBE is a permeable coating and despite a high coating resistance, current could pass directly through the FBE barrier to the underlying steel, developing a high pH environment in the disbondment [Been, Given, Cameron, Worthingham - 2007]. In this case, the water permeation provides a path for CP current to protect the pipe if a disbondment occurs.

Dielectric strength. Wherever a coating is used for corrosion protection, an important characteristic possessed by the coating is its dielectric strength. A coating with an excellent dielectric strength breaks the electric circuit set-up for corrosion to occur. In general, the lower the moisture absorption, the more favorable the dielectric strength of the coating. At the same time, a high dielectric strength of the coating can break the electric circuit for CP current to flow, thus keeping the CP from reaching the substrate metal. Obviously, the effect of a coating's dielectric property is dual. It is good for the coating to be corrosion-resistant, but it shields (at least partially) the applied CP current from reaching the substrate metal.

Resistance to ionic passage. Ionic passage is the transport of chemical ions from external environments through a coating film towards substrate metal. An effective coating should have a good resistance to ionic passage, making itself less conductive and more corrosion-resistant. Even transported ions do not affect corrosion directly, such as sodium ions, which may decrease the coating's dielectric strength, causing it to become more conductive and less corrosion-resistant. Generally, a coating with a high molecular weight and dense molecular structure has a greater resistance to ionic transfer.

2.5.2. Chemical Resistance

The chemical resistance of a coating (and particularly the resins from which it is formulated) is its ability to resist breakdown by the action of chemicals to which it is exposed. Generally, a coating considered to be chemical-resistant must resist salts, acids, and alkali with a wide pH range, and be resistant to organic materials such as diesel oil, gasoline, and similar materials. The resins' chemical resistance to environmental attack is highly selective. For example, epoxies are resistant to alkali and water and less resistant to most acid solutions. In comparison, vinyl coatings are resistant to most acids, alkali, salts, and many solvents such as alcohol, although vinyl is less resistant to organic materials.

Different from the resistance to ionic passage (as discussed above), a coating's chemical resistance refers to its stability in environments containing various chemicals. However, ionic passage through the coating can contribute to the degradation of its chemical resistance. Thus, the two properties are not necessarily the same.

Alkali resistance (i.e., the chemical resistance of a coating in alkaline environments) is very important for a primer in a coating system, especially when the coated structure is under CP. When the CP current penetrates through the coating and reaches the metal surface, electrochemical reactions (i.e., cathodic reduction of water, or oxygen in aerobic environments) generate hydroxyl ions, which elevate the aqueous environment's pH. Thus, a strong alkali is usually developed at locations where CP is enacted. Hydrocarbons like organic coatings generally fail in strongly alkaline electrolytes, resulting in coating disbondment and the spreading of corrosion under the coating. A typical example is pipeline coatings, which are always used with CP. Where the coating is damaged (e.g., pinholes or holidays are generated), CP-driven cathodic disbonding usually occurs on the coating, which is called cathodic disbonding. Therefore, pipeline coatings or the primer contained in a pipeline coating system should be highly resistant to alkali.

2.5.3. Adhesion

Adhesion measures how firmly a coating is bonded to a substrate. Irrespective of its other properties, a coating with a strong adhesion to the substrate retains its integrity much longer than one with other strong characteristics but less adhesion. The property of adhesion is essential for preventing the effect of water on coating performance and service life, and for preventing problems caused by temperature gradients across the coating, osmosis, and electroendosmosis. Moreover, a high adhesion helps to prevent mechanical damage during handling and service, as well as disbondment during service by factors such as soil stress, CP current, etc. The loss of adhesion is usually the first step for corrosion and cracking to occur underneath a coating. Many factors affect coating adhesion, such as substrate chemistry and physics, coating materials and properties, stresses in coating and/or substrate, application procedure, and service environments.

Adhesion is a summation of a wide variety of forces that hold a coating to a structure. It is created by physical, chemical, or mechanical forces that interact where the coating and the substrate meet. A durable coating develops when it comes in contact with the substrate and develops a strong adhesion before curing or drying. Principally, a coating adheres to a metal substrate by three main mechanisms.

Chemical adhesion. Chemical adhesion is created through chemical reactions between a coating and the substrate that generate intermolecular forces or chemical bonds. Intermolecular forces and chemical bonds represent the most effective bonding of all adhesion principles. For example, epoxy molecules bond to a metal substrate with metal hydroxide groups formed through a condensation reaction. Chemical adhesion forms a new chemical compound that firmly joins the coating and substrate.

Polar adhesion. Polar adhesion is a common bonding mechanism, especially for organic coatings. It is created by the attraction of resin molecules to the substrate. Polar adhesion is only possible when the coating and the substrate have the same polarity (i.e., the coating's surface and the substrate's surface must both be either positive or negative). Generally, coatings have lower surface tension than do substrates. The coating wets the substrate with a contact angle less than 90°, which adheres the coating to the substrate.

Mechanical adhesion. Mechanical adhesion is associated with surface roughness or an anchor pattern. Most important to mechanical adhesion is the number of hills and valleys present on the substrate. The substrate pattern increases coating adhesion with the increase in surface area and by the actual roughness. On a microscopic level, any material has a surface composed of valleys and ridges. The surface topography allows the coating to penetrate and fill the valleys, resulting in anchored areas between the coating and the substrate. In addition to substrate roughness and porosity, the coating must have a good "filling power" (wetting of the surface) to penetrate the valleys and pores on the substrate. The coating's "filling power" is directly related to its viscosity.

Some coating types have strong bonds that can be easily measured with various types of "pull off" gauges or peel strength for tapes, shrink sleeves, and extruded coatings. Other coatings, such as wax and viscoelastic coatings typically have cohesive failure at very low peel, yet can be effective coatings when selected for the right environment. Adhesion for each particular type of pipeline coating must be studied and understood, and not just based on the strongest in a particular test.

2.5.4. Flexibility

Flexibility measures the coating's ability to resist mechanical damage (such as expansion and contraction) when it is stretched or bent. A coating's ability to expand and contract along with the substrate is an important quality because a good coating must withstand expansion and contraction from temperature changes without loss of adhesion or cracking. Coating flexibility is affected by temperature, coating thickness, and substrate thickness. Direct exposure to sunlight may change the coating temperature. Therefore, a coating used in a location that is open to air must withstand temperature changes and avoid cracking or loss of adhesion. Generally, thermoplastic coatings (such as PE) possess good flexibility. As temperatures rise, thermoplastic coatings become more plastic and more easily follow substrate expansion and contraction. Thermosetting or cross-linked coatings (such as epoxies) may become brittle with age and unable to contract flexibly with the substrate. This leads to cracking and spalling from the substrate during temperature cycling. Appropriate coating flexibility must be maintained during in-service deformation and temperature changes for the coated structure, but the coating adhesion must not be sacrificed.

Coating flexibility is usually measured in one of three ways (i.e., bending ratio, degree of bend per pipe diameter length (PDL), and strain in percentage of elongation). The bending ratio is the maximum ratio of the bending radius to pipe diameter without coating damage. The degree PDL is the maximum bending angle over the length in one diameter of the pipe. The bend strain measures the coating's maximum possible expansion as a percentage of the elongation. Coatings adhered to substrates are elongated when the substrate bends during installation or in service.

2.5.5. Thickness

A coating's thickness is an important factor in determining its service life and cost. To protect a substrate from corrosion attack and mechanical damage, a certain coating thickness is a must. The physical thickness can actually improve many other coating properties. Generally, the thicker the coating, the better protection it can offer. However, a very thick coating may not always be helpful for adhesion, cracking resistance, and disbondment tendency. A coating's internal stress increases proportionally to coating thickness.

Two terms are used to evaluate the coating thickness (i.e., wet-film thickness and dry-film thickness). Wet-film thickness measurement is an in-process check on the amount of coating being applied. It is primarily used to find potential problems before the coating fully dries. The dry-film thickness inspection measures the thickness of cured solid coating film and gives the final product thickness.

2.5.6. Abrasion Resistance

For ships, helicopter decks, offshore platforms, and similar areas where coatings are subject to the movement of heavy loads and/or to damage by tools and equipment, a coating should be tough, hard, resistant to shock, and adhesive even when used in harsh environments. Abrasion resistance is a coating's ability to withstand mechanical action like rubbing, scraping, or erosion, which progressively removes a coating from a surface. An abrasion-resistant coating maintains its original appearance and structure, and has a longer life especially when the fine-particle impingement type abrasion is present in applications (such as sand abrasion).

It is a common misunderstanding that a coating's hardness is a measure of its abrasion resistance. A coating's adhesion to the substrate plays a major role in abrasion resistance. For example, abrasion-resistant organic polyurethane coatings have a good resistance to impact, scouring, and abrasion. The inorganic zinc coatings, due to their strong adhesion to the steel substrate, are outstanding coatings when applied to barge decks and ship bottom and tops.

2.5.7. Weather Resistance

When coated structures are under direct exposure to sunlight and atmospheric environments, the coating should be weather-resistant. Weather resistance requires the combination of a series of different properties. A weather-resistant coating must withstand the sun's rays; rain, snow, hail, and dew; freezing and thawing; chemical fumes, dusts, and particulate fallout; and continuous dry and wet cycles. To be weather-resistant, a coating must resist these conditions, without excessive chalking, cracking, blistering, loss of adhesion, or substantial color change.

2.5.8. Resistance to Microorganisms

For coated structures where microorganisms are potentially present, the microbial effect on coating properties and performance cannot be ignored. Bacteria settle and accumulate in any dirt present on the coating surface, detracting from coating appearance. Moreover, they attack the coating itself and form colonies or areas that may then be penetrated by corrosion. Catastrophic coating failures can occur due to biological activity. For example, a polyamide coating is vulnerable to biological attack and becomes a food source for bacteria [Munger and Vincent, 1999].

The susceptibility of coatings to biological activity can be controlled with biocides and fungicides added to the coating. Generally, coatings without oils or hydrocarbon byproducts are more resistant to bacteria and fungus growth.

Coatings containing organic sulfides are often subject to breakdown by anaerobic bacteria (e.g., sulfate reducing bacteria (SRB) are present in soil where coated infrastructures like pipelines are in service). SRB can also accelerate metal corrosion through a number of mechanisms [Little and Lee, 2007]. Differential electrolyte concentration cells can be created on the metal surface. Corrosive environments can also be generated due to life cycle and decomposition products. The SRB act as either anode or cathode depolarizers. Moreover, metabolism products (i.e., sulfides) can react directly with metal to form metal sulfides.

External corrosion to buried or immersed cathodically protected pipelines caused by bacteria is typically under disbonded, CP shielding coatings, or bacteria that affect external pipelines live in acidic environments and do not survive in the alkaline environment where adequate CP is available. Under disbonded CP shielding coatings, the environment typically is not affected by the CP enough to prevent the acidic environment. Therefore, bacteria survive and cause corrosion issues.

2.5.9. Resistance to Cathodic Disbonding

Cathodic disbonding, a type of coating failure related to electroendosmosis, is the loss of adhesion between a coating and a metal substrate due to cathodic reductive reactions that take place on the metal surface. CP is applied to prevent metal corrosion, and the applied CP current penetrates the coating and reaches the underlying metal surface. Electrochemical cathodic reaction of water or dissolved oxygen is driven by cathodic polarization for generating hydroxyl ions, which elevates the local solution pH. Since the polymeric coating is vulnerable to alkaline attack, disbonding occurs at the interface between metal and coating.

For pipeline coatings, the recommended CP potential by NACE (National Association of Corrosion Engineers) International is -0.85 V (copper sulfate electrode, CSE). The coating withstands CP of approximately -1.00 V(CSE) without the occurrence of cathodic disbonding. The negative potential of

-1.10 V(CSE) or more can create conditions for cathodic disbonding, depending on the coating and its thickness, dielectric strength, and water resistance [Munger and Vincent, 1999]. The coating must have a low moisture vapor transfer rate and very high adhesion to resist cathodic disbonding.

Cathodic disbonding usually occurs at defects in the coating. The CP current passes through the coating defect and the amount of current flow depends on the size of the defect. With an increase in defect size, the CP current drives the coating disbondment to extend away from the defect, causing an increased disbonding distance. The depth of the coating defect is also a critical parameter affecting cathodic disbonding. Generally, a deep, narrow defect keeps CP current from reaching the defect bottom and avoids cathodic disbonding. However, local corrosion may occur at the bottom of the defect.

The adhesion of the pipeline coatings must be able to withstand the alkaline environment created by the CP. Some coating types will not stay adhered in an alkaline environment. This is the reason cathodic disbondment testing must be performed to determine which coatings will stay adhered in this environment.

2.5.10. Resistance to Soil Stress

Soil stress is particularly important for underground pipeline coatings. Geotechnical factors such as unstable slopes or earthquakes cause significant ground movement. Soil expands and contracts with temperature changes and varied water contents. Geotechnical and environmental factors generate soil stress, which causes primarily longitudinal stress to the coating. Generally, high-clay soils are associated with high soil stress, as clay soil moves substantially and pulls the coating away from the pipe or structure, creating cracks, voids, or thin spots.

Pipeline coatings are also damaged during construction by poor backfill procedures. Rocks can impact the coating, and non-uniform backfill pressure can break the coating. Furthermore, plant roots can either penetrate the coating or surround the pipe and grow to create sufficient pressure that causes coating to flow. Therefore, coatings used under soil-stress conditions must have strong adhesion, high impact resistance, a low tendency to creep or move under pressure, and the ability to resist the abrasion caused by ground movement.

2.5.11. Resistance to Extreme Temperatures

Coatings are designed for use under conditions with moderate (or normal) temperatures and temperature cycles. Coated structures used in extremely cold weather (such as pipelines constructed in Arctic and sub-Arctic regions) benefit from three characteristics (i.e., adhesion, shrinkage, and brittleness [Munger and Vincent, 1999]). The loss of adhesion is a tendency under cold conditions, especially when temperature changes from ambient temperate to a very low value. Almost all organic material becomes more brittle at low temperatures and shatter upon impact. Shrinkage is also a factor in cold weather and affects both adhesion and brittleness. A coating should be highly adhesive to resist extremely cold conditions. It should also be resilient, retaining a certain plasticity at cold temperatures. Coatings that possess these properties as well as sufficient flexibility would not experience a high amount of shrinkage.

2.5.12. Resistance to Environmental Stress Cracking

Environmental stress cracking (ESC) is a phenomenon that affects polymeric resin, which degrades under the synergism of environmental factors and stress. The degradation occurs as cracking and plastic facture of the material. During ESC, environmental chemicals do not directly attack or degrade polymer molecules. Instead, the chemicals penetrate into the coating's molecular structure and interfere with the intermolecular forces that bond polymer chains and lead to molecular disentanglement. Generally, ESC failure processes include fluid (chemical) absorption, plasticization, craze initiation, crack growth, and fracture. Since ESC depends on diffusion of the chemicals into the polymer structure, the rate of fluid absorption is a critical parameter in both crack initiation and propagation. The more rapidly the chemicals are absorbed, the faster the polymer is subject to crazing and subsequent failure. Compared to creep, ESC accelerates with chemicals and results in far less time for crack initiation, in addition to the increased speed of the crack propagation. Thus, the time to failure for polymeric coatings is shortened remarkably. Moreover, ESC of polymers occurs at a reduced stress or strain level relative to the creep failure in air.

2.6. Coating Selection and Application

The proper design and selection of coatings is integral to a structure's design. Proper coating selection helps to maintain or increase the life of an asset by protecting it from environmental attack. Experience has demonstrated that improper selection and a combination of materials that form a coating could be worse than having no coating at all.

2.6.1. Coating Selection Criteria

The type of coating systems selected for certain purposes requires a number of considerations. Factors that affect pipe coating selection include:

- Field-applied or plant-applied
- Surface preparation requirements
- Previous experience with pipe coating
- Anticipated in-service conditions (e.g., temperature extremes)
- Budget constraints
- Anticipated climate conditions during installation (e.g., humidity, moisture, wind, temperature, bending, and jointing)
- Shipping, handling, and installation methods
- Backfill conditions (i.e., pipe bedding and padding)
- Failure mode of the coating

One of the most important criteria in the design and selection of a coating is the type of environment to which the coating will be exposed. The coating must be able to maintain its properties and remain intact and functional, separating the substrate from the environment over a sufficient time period. Environmental factors (temperature, humidity, chemicals, bacteria, etc.) must be considered because they all affect coating properties and performance.

The substrate on which the coating will be applied is also an important consideration. The substrate's physical and mechanical properties, such as its surface roughness, cleanliness, density, porosity, permeability, and geometry must be understood completely. These properties can affect a coating's chemical stability, flexibility, adhesion to the substrate, and so on. In general, steels or alloys are relatively solid, continuous, and dense and ensure a uniform adhesion of the coating to the whole surface. Moreover, most coatings are compatible with steels.

Another consideration involves the potential environmental regulations that restrict the use of certain coatings or materials in specific geographic areas or for some purposes. Typical examples include the restricted use of abrasive blasting and solvent emissions around potable water or food environments.

2.6.2. Storage and Handling

Storage and handling can damage coated pipes. Special precautions are required during storage and handling. Typically, padded equipment is used to handle coated pipes, and proper end hooks are used for handling coated pipes to minimize coating damage. Bunkers on trucks used to transport coated pipes are normally padded. Since point loading on the coated pipe can damage the coating, restrictions on height and the number of coated pipe layers are common. Rope or spacers separate pipes and minimize coating damage. Moreover, the type of the coating must be understood to ensure that proper storage and handling techniques are used.

Frequently, coating material is placed in the back of a truck and left out in the cold or hot, rain or snow, dust, and other environmental factors that affect coating integrity and performance. Field-applied coating materials and related products must be kept at the manufacturers' recommended storage temperatures and protected from other detrimental environmental conditions to ensure that coatings perform properly during and after installation.

2.6.3. Coating Application

Surface preparation. The objective of surface preparation is to create a proper adhesion between a coating and the substrate. To evaluate if the substrate surface is properly prepared for a coating, be sure that the weakest area across the coating system is within the coating layers rather than at the interface between the coating and the substrate. In other words, if the coating fails, the failure should occur inside the coating (resulting in a cohesive failure), rather than at the substrate/coating interface, which would cause adhesive failure. To generate maximum coating adhesion, any extraneous or loose material from the substrate must be removed. Chemically bonded scales, oxide films, and similar surface reaction products must be eliminated. Moreover, a certain surface roughness should be maintained to keep a surface anchor pattern.

Generally, each type of coating has a specific surface preparation requirement that must be followed. For all coatings, the surface of substrates must be clean and dry. This is often accomplished with power tools, but for best performance, abrasive blasting is required. Some coatings require a primer on the prepared surface prior to application of the coating. Primer application must be done in accordance with the manufacturer's specifications, and must be allowed to cure to the specified consistency before the outer coating is applied. Most tape coatings require primers for field application.

Coating the irregularities. The general principle for a coating application to maintain its structural integrity is to keep the structure as simple as possible and to reduce the coated area until it is the smallest area possible. Some common practices for improved coating applications include reduced area and geometry complications, elimination of all overlaps and riveted or bolted points, reduced sharp edges, corners, and rough areas, etc. When coating the interior of facilities, reduce the area exposed to any corrosive substance. If coating is to be applied to a facility's exterior, the facility is preferred to have a flat, cylindrical, and smooth surface, as well as surfaces joined by continuous welding with minimal overlapping joints, rather than joined or bolted connections.

Coatings applied to local irregularities with geometries like angles, channels, H-beams, and I-beams usually experience problems. Angles always have a thinner coating, a large surface tension, and a high internal force, all of which cause poor coating adhesion to the substrate and support the generation of cracks in the coating. Similarly, an interior corner is also an area of difficulty and a danger point. Surface preparation is not easy to conduct, and moisture always accumulates locally. Coatings fail first in areas with sharp edges. Increasing coating thickness at the edges may prevent the local coating failure. Rivets and bolts are other areas where coatings preferentially fail. Double-coating all riveted areas in the same manner as sharp edges would be helpful. From a coating application standpoint, a welded joint is highly preferable to a bolted or riveted joint. Welds should be checked for rough areas, undercut areas, and areas that retain weld slag. Rough welds should be ground to have a smooth contour and all surface imperfections are eliminated.

Coating application. Coatings can be applied to structures in a plant or in the field. They are also commonly applied to a surface during normal operation or during a facility shutdown for repair or maintenance. Coating applications under different situations have specific considerations, including operating temperature, surface cleanliness, applicable methods permissible during operation, personnel present during operation, and safety.

The time and cost required to apply a coating system must be considered. The cost of coating is often only a small part of the total cost for the coating application program. This program includes an environmental evaluation and coating application preparation, surface preparation, coating application, post-treatment, working time, and materials. Depending on coating type, the required preparation and post-treatment (such as drying time) should be considered as they affect the total time of the coating application program.

Assessing and determining when and where a coating is to be applied is also a major consideration. For pipelines, coatings can be applied during construction of a new pipeline project, or as a part of a maintenance or repair program. Generally, new construction projects require complete coating system applications. Newly constructed steel pipes can have coating applied either in a plant or in the field, depending on requirements. For maintenance or repair work, it is not easy to complete surface preparation of pipelines in the field due to either environmental limitations or equipment unavailability. In general, a coating that does not require perfect surface preparation is desirable. Moreover, the compatibility between a newly applied coating and any existing coating must be evaluated.

2.7. Standards for Coating Testing

2.7.1. Primary Standard-establishing Organizations

Pipelines often become international projects, requiring close collaborations and synchronization between participating parties from different territories and countries. Standards have been established by relevant organizations that everyone within the industry can understand and communicate. With pipeline coatings, organizations that develop standards include the American Society for Testing and Materials (ASTM) International, the Canadian Standards Association (CSA), NACE International, the Society for Protective Coatings (SSPC), and American Water Works Association (AWWA).

ASTM International. ASTM International was formed in 1898 to address frequent structure failures experienced in the growing railroad industry. It established standards for steels to be used in railway construction at that time. Currently, the organization has developed more than 13,000 standards that are used worldwide. Generally, ASTM standards are divided into six categories (i.e., standard specification, standard testing method, standard practice, standard guide, standard classification, and terminology standard).

CSA. CSA was formed in 1999 as a global testing and certification organization for assessing product qualities in areas including electrical, mechanical, gas, and plumbing. CSA standards are accepted worldwide and are widely used in North America. Now CSA International certifies products in the following categories: gas equipment, construction products and materials, life science, electrical and electronics, communications, and energy.

NACE International. NACE International, previously named National Association of Corrosion Engineers, was established in 1943 and is specific to corrosion-related industries. NACE's main focuses include CP, coatings, and material selection. NACE has established many standards for practices, test methods, and material requirements for industrial use. NACE standards are updated on a five-year basis.

SSPC. SSPC was founded in 1950, and was initially named as the Steel Structure Painting Council. SSPC focuses on the protection and preservation of concrete, steels, and other industrial and marine structures and their surfaces. SSPC provides information to the coating industry regarding surface preparation, coating selection and application, environmental regulations, and health and safety issues.

AWWA. AWWA is a non-profit organization formed in 1881 to provide standards and regulations for water management. The AWWA Standard Program is an internationally recognized source. Currently there are 150 AWWA standards that cover a wide range of areas such as filtration materials, treatment chemicals, disinfection practices, meters and valves, storage tanks, pumps, and pipe fittings for various types of pipes. The AWWA does not directly relate to the oil and gas pipeline industry, but it contains standards that apply to pipeline coatings.

2.7.2. Important Standardized Testing Methods for Pipeline Coatings

Generally, standards developed for design, selection, testing, and maintenance of coatings (including pipeline coatings) are divided into three types:

- Type I standards for design coatings to be used under specific environments (e.g., coated pipelines buried in moist soils)
- Type II standards for the evaluation and selection of specific types of coatings (e.g., FBE coating applied to steel pipes)
- Type III standards for testing coating properties (e.g., adhesion of the coating to substrate materials)

For Type I standards, the major establishing organizations are NACE and CSA. For Type II standards, predominant providers include NACE, AWWA and CSA. ASTM is the main organization developing Type III standards. Some commonly used ASTM standard methods used for coating testing are detailed in the following sections.

Standard method for measuring adhesion of pipeline coatings [ASTM, 2009]. As previously stated, adhesion is one of the most important properties a coating possesses to be functional. Ideally, a non-permeable coating that is perfectly adhered to a substrate shall require no further measures to protect the substrate from corrosion attack. In reality, the coating adhesion cannot be maintained at a sufficiently firm condition over long-term service. Evaluation of a coating's adhesion to a substrate is critical for predicting coating performance in the field, and also for obtaining knowledge in the design process for coating optimization. Table 2-2 describes two ASTM standardized methods for testing coating adhesion.

Table 2-2. Standard Testing Method for Adhesion of Pipeline Coatings [ASTM, 2009]

Standard	ASTM D3359-09
Skills prerequisite	No
Procedure (method A)	· Select an area on the test specimen, i.e., coated steel specimen, free of imperfections as the test area. Clean the area if needed. · Make two 40 mm cuts on the test area. The two cuts intersect each other at 30^0 to 45^0. Each cut must cut through the coating onto the substrate in a continuous motion. · Inspect using reflection of light to ensure that the coating has been fully penetrated. If not, repeat above steps at a new test area or specimen. · Carefully cut 75 mm of tape. · Line the center of the tape with the intersection of the cuts, and the length of the tape goes along the smaller angles of the cuts. Press down firmly with finger and run the eraser along the length of the tape. The color shown through the transparent tape is a good indication of good contact (shaper colors means better contact). · Within 60-120 seconds, pull tape off the specimen in one rapid and continuous motion. · Inspect the X-cut area and rate it based on the scale table included in the standard. · Clean cutting tool and repeat process on 2-3 more spots on the test specimen.

Procedure (method B)	· Select an area on the test specimen that is free of imperfections as the test area. Clean the area if needed. · Make 11 parallel cuts of 20 mm long for each, with the spacing between cuts specified in the standard. Each cut must cut through the coating and onto the substrate surface. · Brush the cut area with a soft brush to remove any detached flakes. · Clean cutting tool and make 11 more parallel cuts at 90^0 to the original cuts. Use the same cut geometry as stated above. · Brush the cut area with a soft brush to remove any detached flakes. · Examine the cut area by light reflections to ensure full penetration of coating. If metal surface is not reached, repeat the above steps at a new test area or specimen. · Carefully cut 75 mm of tape. · Tape over the cut grid (ensure maximum coverage) and run eraser along the length of the tape. The color shown through the transparent tape is a good indication of good contact. · Within 60–120 seconds, pull the tape off the specimen in one rapid and continuous motion. · Inspect the grid and rate based on the scale table included in the standard. · Clean cutting tool and repeat process on 2-3 more spots on the test specimen.
Comments	The testing results should be used as relative ratings to rank different types of coating on the same substrate or one type of coating on different substrates. The results should not be taken as absolute adhesion performances.

To follow the adhesion testing standard, results can be affected by the different tapes used. It is recommended that each piece of tape is taken from the same roll to evaluate the adhesion of various coatings.

Standard method for measuring bendability of pipeline coatings [ASTM, 2010a]. Pipelines are usually designed to withstand extreme conditions during service life. One typical example includes big variations in temperature, which may cause coating contraction and expansion. This may result in disbonding and cracking of the coating. Thus, a pipeline coating should possess sufficient flexibility (or bendability) to resist extreme temperature changes. Table 2-3 shows the standard method for testing pipeline coating bendability. A precision statement is not applicable for the testing method, as no quantitative results are obtained.

Table 2-3. Standard Testing Method for Bendability of Pipeline Coatings [ASTM, 2010a]

Standard	**ASTM G10-10**
Skills prerequisite	Certified lab technicians/analysts
Procedure	· Let the test specimen rest at room temperature (20^0 to 30^0C) for sufficient time to ensure a thermal equilibrium. · Measure the coating thickness of the coated specimen using a thickness gauge. · Secure one end of the specimen with a No. 1 clamp and remove threaded handle to allow for specimen clearance during testing, and install the coated specimen in the bending jig as instructed. · Apply a constant and even bending force on the specimen with the lever handle. · Bend the specimen until it is able to comfortably clamp into the No. 2 and No. 3 clamps as specified. · Apply a constant and even bending force on the pipe with the lever handle until the roller rolls off the end of the specimen. · Visually examine the specimen for any disbonded and cracked coating. Use the holiday detector to aid the examination process. Note down all findings such as their size, location and type.
Comments	This method provides a way of investigating relative merits of different coatings.

Standard method for measuring the brittleness temperature of pipeline coatings [ASTM, 2014]. Pipelines operating below the freezing point are subject to brittle failures of the coating. For coatings used in Arctic and sub-Arctic areas, the ability to withstand brittle failures is an important property. Determining the brittleness temperature allows a complete understanding and evaluation of the coating performance in the field. It also guides the coating selection during the design process. Table 2-4 shows the standard testing for determining the temperature at which a coating experiences a brittle failure. The test results may vary with the methods for specimen preparation. For comparable results, use the same technique to prepare the specimens.

Table 2-4. Standard Testing Method for Determining the Brittleness Temperature of Pipeline Coatings [ASTM, 2014]

Standard	ASTM D476-14
Skills prerequisite	Certified lab technicians/analysts
Procedure	· Predefine the temperature at which 50% of specimens are expected to fail using accredited resources or past experience. · Set the specimen bath to the predefined temperature. · Clamp specimens and ensure appropriate tightness with a torque wrench. · Mount the clamp in the testing apparatus and start test cycle. · Remove specimens when the cycle is complete and allow them to be steady to room temperature. · Evaluate specimens for cracking, and record the percentage of specimens that contain cracks. · Change the bath temperature at 5°C increments to determine the temperature at which none of the specimens show cracks. Record the percentage of specimens that contain cracks under each temperature increment. · Go back to the predefined temperature and change the bath temperature in 5°C decrements to determine the temperature at which all of the specimens show cracks. Record the percentage of specimens that contain cracks under each temperature increment. · Plot a temperature vs. percentage of failure graph using the data points. · Draw a best-fit line, and the temperature at which 50% of the specimens fail is the brittleness temperature.
Comments	This test is useful for evaluation of the relative performance in similar scenarios, but it does not indicate the minimum temperature at which the coating can be used.

Standard method for testing impact resistance of pipeline coatings [ASTM, 2013]. The majority of transmission pipelines are buried. Excavation and later reinstallation of soil over the pipeline are required. During construction, the coating may suffer from inevitable damage. To minimize damages, the resistance of the coating to mechanical impact needs to be assessed. Table 4-5 shows the standard test for evaluating the impact resistance of coatings. There is no information for the precision and repeatability of the test results, as indicated in the standard.

Table 2-5. Standard Testing Method for Impact Resistance of Pipeline Coatings [ASTM, 2013]

Standard	ASTM G13/G13M-13
Skills prerequisite	Certified lab technicians/analysts
Procedure	· Follow the installation procedure to install the test specimen. · Fill a bucket with 35 lb (16 kg) of stones. · Drop stones all at once onto the specimen. · Remove specimen after each drop and assess coating condition with a holiday detector and thickness gauge. · Repeat until the specimen has section(s) fully exposed of the bare steel substrate. · After 5 buckets of stones have been dropped, replace the stones with a new batch. · Perform test up to 10 buckets (terminate test if no damage is discovered after 10 drops). · Record the number of buckets dropped, which will be the measurement used for evaluation of the impact resistance of the coating.
Comments	Perform test at room temperature (21°C to 25°C).

Standard method for testing water resistance of pipeline coatings [ASTM, 2007]. The deterioration and service life of a coating is closely related to its capacity to resist water permeation. Water resistance is perhaps the most important property a coating possesses. Generally, the larger the water resistance, the better the coating is. If water penetrates through the coating, it can reach the bare metal substrate under disbonded coating and initiate corrosion and coating degradation. Table 2-6 shows the standard method for testing water resistance of coatings. The difference between results, obtained from tests conducted consecutively on the same specimen under identical conditions, may normally be expected to not exceed ±5%.

Table 2-6. Standard Method for Testing Water Resistance of Pipeline Coatings [ASTM, 2007]

Standard	ASTM G9-07
Skills prerequisite	Certified lab technicians/analysts
Procedure	· Prepare test specimen as described in the standard. · Assemble test cell as instructed, and connect the test cell with a capacitance meter. · Measure the capacitance of the coating (C_c), the initial capacitance (C_0), and the dissipation factor (D_F) of the specimen. · Make periodic measurements of the capacitance (C_c) and dissipation factor (D_F) of the test specimen. · Calculate dielectric constant of the coating, and the apparent depth of penetration.
Comments	The water penetration test provides the means for monitoring the passage of moisture through a coating by means of changes in its dielectric constant. The data will reflect a rate of deterioration and provide information for establishing the optimal coating thickness.

Standard method for testing cathodic disbonding of pipeline coatings [ASTM, 1996; ASTM, 2011]. As stated, a good coating can fully protect the pipeline from corrosion attack. However, it is difficult, if not impossible, to achieve perfect performance in reality, especially over a long time. Thus, CP is always applied on coated pipelines as a back-up measure for corrosion protection. While the CP prevents corrosion of the target structure (such as pipelines) by cathodically polarizing the metal (i.e., pipeline steel), it can also weaken the coating's adhesion to the substrate metal. This causes disbondment of the coating by generation of an alkali environment at the metal/coating interface. Due to the dual nature of CP, it is important to test the coating's ability to resist disbonding caused by CP. Table 2-7 describes the procedure for determining coating characteristics to resist cathodic disbonding where the specimen is under CP.

Table 2-7. Standard Method for Testing Cathodic Disbonding of Pipeline Coatings [ASTM, 1996; ASTM, 2011]

Standard	ASTM G8-96 & ASTM G42-11
Skills prerequisite	Certified lab technicians/analyst
Procedure (ASTM G8-96)	· Set up the equipment as described in the standard. · Immerse the sample in the electrolyte and position the intentionally opened holiday, named intentional holiday, to face the anode. Mark the immersion level and maintain by daily additions of potable water. Keep the electrolyte temperature of 21 to 25°. · Measure the potential between the test specimen and a reference electrode. It should be 1.45 V to - 1.55 V (copper/copper sulfate, CCS). · Perform the test for 30 days. · Take out the specimen from the vessel, wash it with warm tap water and dry it immediately. · Drill a new reference holiday in the coating in an area that was not immersed. · Make radial 45° cuts through the coating intersecting at both reference holidays with a sharp-bladed knife. · Lift the coating at the holidays using a thin-bladed knife. Use the bond at the reference holiday for judging the quality of the bonding at the intentional holiday. Measure and record the total area of disbonded coating at the intentional holiday.
Procedure (ASTM G42-11)	· Combine NaCl, Na_2SO_4, Na_2CO_3 to create the electrolyte solution. · Immerse the test specimen in the electrolyte and connect it to the anode. · Heat and maintain the electrolyte temperature at no less than 60 °C. · Determine and record the resistivity and pH of the electrolyte at the beginning and end of the test period. · Standard duration of the test period is 30 days.
Comments	The two standards are similar, with the only difference in that ASTM G42-11 measures the cathodic disbondment at an elevated temperature. It is important to mark the correct immersion level on the exterior of the test vessel and maintain such level by daily additions of preheated, or distilled water as required. The apparent resistance of the pipe after immersion shall be stable and not less than 1000 MΩ.

The surface condition of the specimen and the coating application affect the precision of the results. Specimens should be taken from adjacent segments on the same coated pipe. Specimens taken from different lengths of the pipe may represent differing process conditions, and generate erroneous results.

Standard method for testing tensile properties of pipeline coatings [ASTM, 2010b]. Buried pipelines may experience significant soil stress or strain due to ground movement at unstable slopes or in earthquake zones. For coating to be effective, it must have the required mechanical properties for resisting extensive soil stress and strain. Table 2-8 details the standard for assessing tensile properties of coating materials.

Table 2-8. Standard Method for Testing Tensile Properties of Pipeline Coatings [ASTM, 2010b]

Standard	ASTM D638-10
Skills prerequisite	Certified lab technicians/analysts
Procedure	· Clamp the coating specimen tightly by the ends (it is very important to ensure a secure grip) on a materials testing machine enabling tensile testing. · Set machine speed according to the specifications described in the standard. · Save results obtained from test. · Obtain material's tensile strength, elongation (% elongation, % elongation at yield, % elongation at break and nominal strain), elastic modulus and secant modulus as needed.
Comments	Load indicator accuracy needs to be determined, and if required, calibration needs to be performed daily.

Many variables, such as specimen preparation, surface treatment, testing speed, and service environment must be controlled in the test for comparative results. Testing precision varies with each variable. Thus, repeatability studies have been conducted to ensure data accuracy and validity.

2.8. References

ASTM (1996) Standard Test Methods for Cathodic Disbonding of Pipeline Coatings, ASTM G8-96, West Conshohocken, PA, USA.

ASTM (2007) Standard Test Method for Water Penetration into Pipeline Coatings, ASTM G9-07, West Conshohocken, PA, USA.

ASTM (2009) Standard Test Method for Measuring Adhesion by Tape Test, ASTM D3359-09, West Conshohocken, PA, USA.

ASTM (2010a) Standard Test Method for Specific Bendability of Pipeline Coatings, ASTM G10-10, West Conshohocken, PA, USA.

ASTM (2010b) Standard Test Method for Tensile Properties of Plastics, ASTM D638-10, West Conshohocken, PA, USA.

ASTM (2011) Standard Test Method for Cathodic Disbonding of Pipeline Coatings Subjected to Elevated Temperatures, ASTM G42-11, West Conshohocken, PA, USA.

ASTM (2013) Standard Test Method for Impact Resistance of Pipeline Coatings (Limestone Drop Test), ASTM G13, West Conshohocken, PA, USA.

ASTM (2014) Standard Test Method for Brittleness Temperature of Plastics and Elastomers by Impact, ASTM D746-14, West Conshohocken, PA, USA.

Bedard, Y., Riedl, B. (1990) Synthesis of a phenol-formaldehyde thermosetting polymer, *J. Chem. Edu.* 67, 977-978.

J. Been, R. Given, K. Ikeda-Cameron and R. Worthingham; "Investigating Coating Performance"; Pipeline and Gas Technology, April 2007, page 36.

Bohannon, J. (2005) 'Smart coatings' research shows the virtues of superficiality, *Corros.* 309, 376-377.

Buchanan, R. (2003) Pipeline coatings & joint protection: A brief history, conventional thinking & new technologies, RIO Pipeline 2003 Conference, Rio de Janeiro, Brazil.

Craver, K.J., Tess, R.W. (1975) *Applied Polymer Science*, Organic Coatings and Plastics Division, Amer. Chem. Soc. p. 518.

Encyclopedia Britannica (1974) *Paints, Varnishes and Allied Products*, Vol. 13, p. 886.

Francis, R.A. (2013) Inorganic zinc silicates – The early days, Part I: Invention and early applications, In: *Inorganic Zinc Coatings: History, Chemistry, Properties, Applications and Alternatives*, R.A. Francis, Editor, Australasian Corrosion Association, p. 7-17.

Guan, S.W. (2001) 100% solids polyurethane coatings technology for corrosion protection in water and wastewater systems, the 9th Mideast Corros. Conf., Manama, Bahrain.

Knop, A., Pilato, L.A. (1986) *Phenolic Resins*, Springer, Berlin, Germany.

Kumar, A., Stephenson, L.D. (2004) Self-healing coatings using microcapsules and nanocapsules, Corrosion'2004, paper no. 04278, NACE, Houston, USA.

Kumar, A., Stephenson, L.D., Murray, J.N. (2006) Self-healing coatings for steel, *Prog. Organ. Coat.* 55, 244-253.

Lamaka, V., Zheludkevich, M.L., Yasakau, K.A., Serra, R., Poznyak, S.K., Ferreira, M.G.S., Nanoporous titania interlayer as reservoir of corrosion inhibitors for coatings with self-healing ability, *Prog. Organ. Coat.* 58, 127-135.

Little, B.J., Lee, J.S. (2007) *Microbiologically Influenced Corrosion*, Wiley, Hoboken, NJ, USA.

Munger, C.G., Vincent, L.D. (1999) *Corrosion Prevention by Protective Coatings*, Ed. 2, NACE Publishing, Houston, TX, USA.

Niu, L., Cheng, Y.F. (2008) Development of innovative coating technology for pipeline operation crossing the permafrost terrain, *Constr. Build. Mater.* 22, 417-422.

Norsworthy, R. (2009) Coatings used in conjunction with cathodic protection – shielding vs non-shielding pipeline coatings, CORROSION 2009, Paper 09043

PSG Web Source, *A Brief History of Paint & Coatings*, http://www.psgdover.com/en/paint-coatings-resources/timeline-history.

Schweitzer, P.A. (2005) *Paint and Coatings: Applications and Corrosion Resistance*, CRC Press, USA.

World Book Encyclopedia (1978) *Paint*, Vol. 15, p. 24.

Zaki, A. (2006) *Principles of Corrosion Engineering and Corrosion Control*, Elsevier, Amsterdam, The Netherland.

3 Development of Pipeline Coatings

Over the past half-century, coating technology has experienced revolutionary innovations in pipeline protection from corrosion attack. A wide variety of pipeline coatings has emerged with continuously improved corrosion protection ability. This includes material technology innovations from coal tar to asphalt, sintered polyethylene (PE) and fusion-bonded epoxy (FBE), and the development of new systems, from single-layer coatings to two- and three-layer systems as well as composite coatings.

Uses of pipeline coatings in different countries and regions are based on regional recognition and preference of the coating's specific properties. For example, in North America, the resistance to cathodic disbondment in epoxy-based coatings is the favorite property, while European users emphasize the importance of the coating barrier's blocking effect on external environments, resulting in extensive uses of thick PE-based coatings.

The use of coating products also depends on a variety of other factors including cost, applicability, durability, repairability, environmental concerns, etc. Figure 3.1 shows the global development of coating products in Canada since the 1940s; FBE was first used in 1960 [Shaw Pipe, 2012]. Early products like coal-tar enamel and asphalt are no longer used to coat newly constructed pipelines due to generally poor field experiences and health hazards. Solid film-backed PE tapes are also declining in use because of their poor adhesion, soil stress issues, and the associated external corrosion and stress-corrosion cracking (SCC) occurrence. FBE and FBE-based, multi-layered products where FBE is used as a primer are presently the dominant anti-corrosion coatings applied to most pipelines, especially in North America. FBE possesses various properties of superior coatings, including anti-corrosion performance and cathodic-protection (CP) compatibility. FBE is not likely to sustain mechanical damage, and its durability leads to the development of FBE-based, dual- and triple-layered coating systems, including two-layer polyethylene (2LPE), three-layer polyolefin (3LPO), and 3LPE or three-layer polypropylene (3LPP). Currently, the single-layer FBE is more popular in North America and the UK, while dual-layer FBE products are more popular in Australia. The 3LPO coatings dominate the European and Chinese pipeline coating markets [Guan, 2011].

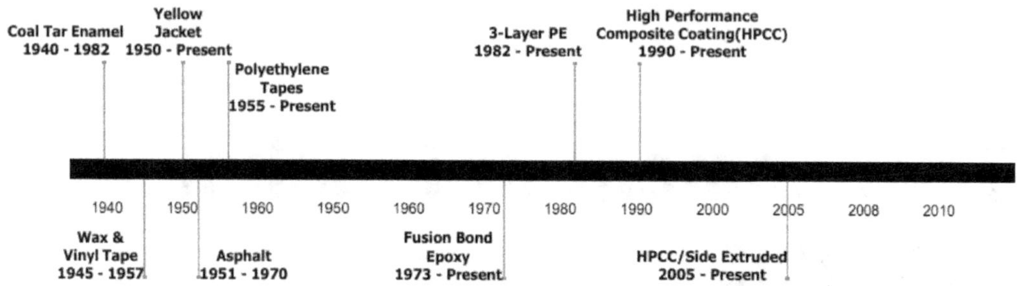

3-1. FBE has been used in Canada since 1960.
Figure 3-1 represents Shaw's first use of it. [Shaw Pipe, 2012].

3.1. Plant-applied Pipeline Coatings

Plant-applied coatings are intended to provide the most in combinations of mechanical properties and corrosion protection, although the method of application is not the primary consideration. Plant coatings can be applied on substrate, independent of environmental influences and human factors. Coatings applied in the field must be comparable to mechanical properties and corrosion resistance offered by the plant-applied coating, ensuring similar application and performance under conditions encountered on-site.

3.1.1. Coal Tar

The good film-forming and waterproof properties of coal-tar products have been known for centuries, but it has only been during the last century that they have been developed for pipeline use. Coal tar was the first type of material used for pipeline coatings beginning in the 1940s. Coal-tar coatings were also some of the first pipeline coatings to be applied in a plant successfully. Due to the restrictive environmental and health standards required by agencies like the Occupational Safety and Health Administration, the Environmental Protection Agency, and the Food and Drug Administration, coal-tar coating declined in use in the late 1970s, when new coating materials like PE with improved properties became available. Today, coal-tar coatings are still in use in some countries like India [Schad, 2013]. Some coal-tar plants still exist in North America; most coal-tar coating is applied to pipe used in the water industry. In addition to its use as a coating, coal tar can be distilled to produce refined tar and pitch, used extensively for road construction, roofing, and waterproofing purposes (as well as for other industrial applications).

Coal-tar pitch, which forms the basis for coal-tar enamel, consists of polynuclear aromatic hydrocarbons and heterocyclic three- to six-ringed compounds. These stable molecules are formed during coking operations at about 1,300°C. During manufacturing, liquid coal tar disrupts the coal's secondary valence forces and penetrates its plate-like structure as it dissolves. The fillers and coal add flexibility and strength to the product. The strong molecular bonds provide an exceptional resistance to water penetration.

Normally, coal-tar pitch is sensitive to temperature changes. It can become hard and brittle at low temperatures, resulting in crevices and cracks and with a significant decrease in adhesion to the steel surface. Coal-tar coatings tend to soften and flow at high temperatures. Recent improvements have been made to reduce coal-tar coating sensitivity to temperature. For example, a method was used by which dissolution of bituminous coal in coal tar reduced coal-tar susceptibility to temperature changes.

3.1.1.1. Coal-tar Enamel

Coal-tar enamel (CTE) is a polymer-based coating produced from plasticized coal-tar pitch, coal, and distillates. Inert fillers are added to provide desired properties of the system. A filled coal-tar pitch has a higher softening temperature than unfilled material and has a lower tendency to flow. This is important in tropical countries or for pipes operated at elevated temperatures. Over the years, this coating has been used with a primer, glass fiber or mineral felt reinforcements, and an outer wrap [Romano et al., 2012]. The introduction of glass fiber inner wraps and the application of outer wraps on the coating surface remarkably improved the system's mechanical strength and provided extra protection against soil stress and impact damage during handling and installation.

The CTE system includes four main components: primer, CTE, glass-fiber inner wrap, and glass-fiber outer wrap. When selecting enamel and outer wrap grades, it is crucial to consider performance requirements like service temperatures, local ground conditions, and seasonal variations.

Primer. Two types of primer are available for manufacturing the CTE system (i.e., synthetic primers and epoxy primers). Synthetic primers are based on chlorinated rubber that can be applied by spray, brush, or roller and become dry within 5-15 minutes. Epoxy primers are two-component materials and are applied by multi-component spray equipment on warm pipes (100-140°F). Both primers are compatible with all different grades of enamel.

CTE. There are four grades of CTE coating, all fully compatible with one another but designed for use under different service conditions and temperatures. Each grade is manufactured from the same raw material and differs only by the ratios used. In general, softer enamels are used at low in-service temperatures; harder enamels are used at high in-service temperatures. It is important to select the correct CTE grade for each pipeline. High-temperature sections, such as those right after compressor stations, may require a hot line grade.

Inner wraps. Inner wraps are resin-bonded glass fiber reinforcements with limited elasticity. These reinforcements add mechanical strength to enamels, which are usually thermoplastic materials. Inner wraps strengthen the coating, inhibit creep and increase impact resistance. For coating thickness below 4 mm, it is recommended that one inner wrap is sufficient; for coating thicknesses of 4 or more mm, two inner wraps are advisable.

Outer wraps. Outer wraps are CTE-impregnated glass-fiber materials. They are porous and allow vapors to escape during application, enabling the hot enamel to permeate and fuse the outer wrap to the surface. These glass-fiber outer wraps have replaced old-fashioned and environmentally unacceptable asbestos felts, which were not porous and often entrapped vapors, leading to voids or detached felt. Selecting the correct grade of outer wraps depends on what mechanical stress the coated pipe encounters during handling, transportation, and installation. Generally, soil stress can generate a maximum force along the pipeline in the 12 o'clock position, which is proportional to the pipe diameter. This would be the required longitudinal strength for outer wraps. All grades of CTE and coal-tar outer wraps are compatible with one another.

CTE coatings can be applied in both plant and field. With plant application, the pipe is abrasive-blasted to remove any rust and mill scale. After a coating of primer is applied, the pipe is rotated. The CTE is heated to approximately 240°C and then poured over the rotating section of the pipe surface [Romano et al., 2012]. A glass-fiber inner wrap is immediately pulled over the coating. A second glass-fiber is applied over the first wrap. A second layer of CTE is then applied. Liquid CTE spreads through the wrap. The coating is then finished with one coat of white wash or a single wrap of Kraft paper to prevent UV degradation of the enamel during storage in direct sunlight, as shown in Figure 3-2. The CTE coating has a minimum thickness of 2.4 mm.

Figure 3-2. Coal tar enamel coated steel pipes in storage (concrete coated).

With in-field application, the pipe is brought to the right-of-way (ROW). After the pipe sections are welded together, a cleaning unit consisting of rotating wire brushes removes mill scale, rust, and welding irregularities. After a primer is applied, the hot, melted coal-tar coating is applied to the pipe with a glass wrap. The coated pipe is finally wrapped with a protective outer wrap [McManus et al., 1966]. Because environmental conditions could not be controlled and because surface preparation was not always ideal, field-applied coal tar had a tendency to fail more often than plant-applied coal tar.

In summary, the CTE coating has an outstanding record for pipeline protection. Technical developments have enabled the system to be applied rapidly and efficiently at costs substantially below competing products. By selection of correct grades, it is reliably used for in-service temperatures between

-30°C and +95°C and is effective up to 115°C under concrete weight coats for subsea pipelines. At low temperatures, precautions should be taken to prevent cracking and disbonding of the coating during field installation.

3.1.1.2. Coal-tar Epoxy Coatings

Coal-tar epoxy coatings are made by blending an epoxy resin with coal-tar resins. The coating possesses the combined properties of water resistance and adhesion owned by coal tar, and the chemical resistance of epoxy improves the performance of the resulting coating [NACE, 2011].

To date, coal-tar epoxy coatings have been used in applications like oil-tank interiors, pipelines, ship hulls, sewage and water treatment plants, and oil platforms. Coal-tar epoxy coatings have the advantages of strong adhesion, good chemical resistance, high-temperature resistance, and good resistance to moisture permeation. The coal-tar epoxy coating is generally used alone, generating a relatively thick film. When cured, the coating develops a hard, slick film, which is difficult to recoat or repair.

3.1.2. Asphalt

Asphalt is a bituminous material obtained either from coal-tar pitches or as a residue from the distillation of petroleum asphalts. Asphalt can be used for a variety of purposes because of its good chemical and weather-resisting properties. These include highway surfacing, airport runways, concrete roads, etc. Asphalt can also be used as paving material and for waterproofing and preserving roofing materials. Asphalt is one of the world's oldest engineering materials and has been used since the beginning of civilization.

While naturally occurring or mined asphalts such as gilsonites exist, the majority of asphalt types are from asphaltic petroleum. The physical and chemical characteristics of asphalt coatings obtained from distilling asphaltic petroleum depend on distillation temperature and various processing conditions, as well as the nature of the crude petroleum to be distilled [Tracton, 2010]. In the 1940s, asphalt began to be introduced in pipeline-coating development.

Asphalt, along with coal tar, was a commonly used pipeline coating in the 1950s. Asphalt coatings have long been one of the most popular types of protective coatings for prevention of corrosion of buried oil and gas pipelines. They were mainly applied in the field and used with CP, primer and outer wrappers. The important performance criterion for evaluating asphaltic coatings is the coating's adhesion to steel substrate in the presence of CP and deformation that resulted from mechanical and soil stress [Alexander and Tarver, 1965]. In addition to a lengthy coating application time, asphalt-coated pipelines also suffered from problems with performance and integrity due to other factors such as poor surface preparation and application by unskillful workers under undesirable weather conditions.

As plant-applied coatings become favorable, asphalt coating gradually yielded to other newly developed coatings. With the development of more advanced bituminous-based coating products (for example, asphalt enamels, where filling materials improve the asphalt coatings' impact and abrasion resistance), asphalt coatings are presently used to coat pipelines. Petroleum asphalt has been used as a protective coating with and without filling materials. When the asphalts are filled, they are termed mastics or enamels.

3.1.2.1. Fillers

Fillers are normally added up to a maximum of about 30% by weight (calculated on the mixture), which is equivalent to about 15 to 20% by volume. A satisfactory filler must have the following characteristics:

- Low water absorption. Certain fine clays are unsuitable.
- Ability to be readily wetted by enamel.
- Finely-ground composition, with particles preferably of laminar shape to prevent settling when the enamel is molten.
- Relatively low specific gravity, so that there is a minimum tendency for the filler to settle out in the melting kettle.

There is an optimum percentage of fillers that help a coating possess the required melting point and toughness. Beyond this optimum percentage, application becomes more difficult and water tightness may be impaired.

3.1.2.2. Asphalt Mastic Coating

Asphalt mastic coatings are asphalt coatings that contain a dense mixture of sands, crushed limestone, and fibers; they are bound together with selected air-blown asphalt. The coating is a thick (12.7 to 16 mm), extruded mastic that results in a seamless corrosion-resistant coating. Extruded asphalt mastic pipe coating has been used for more than 50 years.

Coating selection is based on operating temperature and climatic conditions to obtain maximum flexibility and operating characteristics. The asphalt mastic coating can be designed for operation over a wide temperature range of 40–190° F (4.4–88°C). Precautionary measures should be taken when handling asphalt mastics in freezing temperatures. UV rays should be avoided, especially for coated pipes in storage. Moreover, the asphalt mastic coating is unsuitable for use above ground and in soils contaminated with hydrocarbons.

Hot applied asphalt coatings consist of 75–85% air-blown asphalt and 15–25% of finely divided mineral fillers used mainly to improve coating toughness. A primer is also used to enhance the adhesion of the coating to steel substrate [Alexander and Tarver, 1965].

3.1.2.3. Asphalt Enamel Coating

Asphalt enamels have been successfully used in oil and gas transmission and distribution pipelines for many years with long-term corrosion protection. The majority of today's pipelines are coated with hot-applied plasticized coal tar or petroleum asphalt enamels. Many of today's pipelines are field-coated with hot-applied plasticized coal tar or petroleum asphalt enamels. Compared to coal-tar enamels, asphalt enamel coatings are safer and more environmentally acceptable. Asphalt enamels are often combined with glass fiber or external wraps to provide a good mechanical strength during handling, and to protect the coating from UV rays. However, the asphalt enamel coatings are often subject to cracking and disbonding during field application when temperatures fall below 40°F (4.4°C) [Munger, 1999]. With improved technologies, the coating can be used at minimum and maximum operating temperatures of -40°F (-40°C) to 158°F (70°C), respectively.

The asphalt enamel coating system can be applied to various pipe diameters with varying grades of reinforcement wraps. Figure 3-3 shows the application process of asphalt enamel in the plant [BrederoShaw 1]. The coating is compatible with CP, but over time water can permeate the coating, resulting in coating degradation. If the applied CP current is insufficient, corrosion can occur on pipe steel. Currently, increased awareness of its potential health hazards and its susceptibility to SCC has reduced the use of asphalt enamel coating on pipelines.

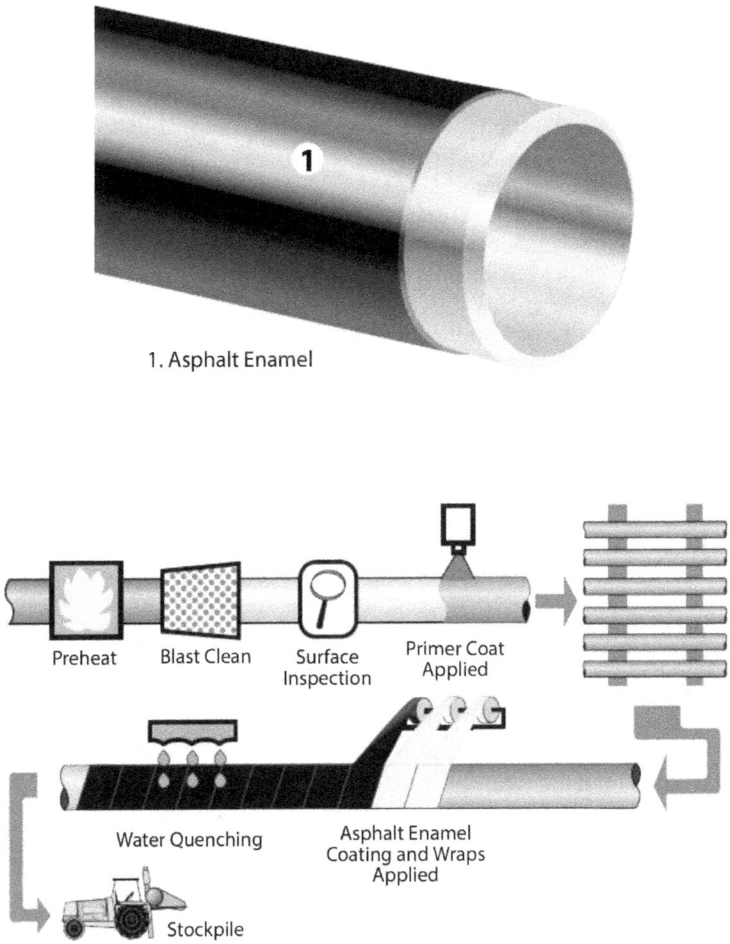

Figure 3-3. Asphalt enamel and its application process in plant [BrederoShaw 1].

3.1.2.4. Comparison between Coal Tar Pitch and Asphalt Coatings

Asphalt and coal-tar pitch are both waterproof materials and resemble one another's physical type. In many circumstances, they are both very effective in preventing the access of water to a coated steel surface. It is often claimed that a coal-tar coating absorbs less water than an asphalt coating, and there is evidence to support this claim. However, some in-practice asphalt enamels have performed as well as the best coal-tar enamels. Coal-tar enamels are believed to adhere better to clean metals than asphaltic enamels (probably because of the presence of polar compounds), but little difference is found in practice under proper pipelining conditions. Asphaltic enamels are easier to apply, as they do not produce such obnoxious fumes and are usually applied at slightly lower temperatures.

3.1.3. Liquid Epoxy Coatings

Liquid coating systems are most often used for maintenance and repair. Rehabilitation coatings for pipelines' and tanks' internal coatings are often liquid coating systems. An advantage of using liquid coatings is that (under most conditions) they do not require extra heat to achieve cure. Some are solvent types, and others are 100% solids. These systems are primarily used on larger-diameter pipes and offer good corrosion resistance at operation temperatures up to 203°F (95°C). These liquid products are usually two-component materials. For example, epoxy coatings are two-part, ambient temperature-cured, 100% solids, thermosetting materials with a base resin and curing agent. Polyurethanes are also two-part, 100% solids coatings, but with a polyisocyanate curing agent and a polyol. They are easily applied in the field or in the plant and provide excellent performance. Typical uses of liquid coating systems include girth welds, valves, fittings, pipes, tanks, ships, etc. Liquid coating systems used with CP are usually only applied in the field.

Generally, epoxies have an amine or a polyamide curing agent. Coal-tar epoxies have coal-tar pitch added to the epoxy resin. A coal-tar epoxy cured with a low-molecular-weight amine is especially resistant to alkaline environments (for example, those generated on a cathodically protected structure). Some coal-tar epoxies become brittle when exposed to sunlight.

The application process for liquid coatings consists of several simple steps (i.e., cleaning substrate surface, applying the liquid coating, and allowing the coating to cure).

Substrate cleaning. Foreign materials (e.g., old coating, dirt, oil and grease) and weld spatter should be removed, and sharp edges should be ground to a radius of approximately 0.12 inch (3 mm). The area should be blast cleaned to an acceptable standard as required in NACE No. 2/SSPC-SP 10, Near-White Blast Cleaning [NACE/SSPC, 2010]. The blast cleaning establishes an anchor pattern or surface roughness, with typical specifications for a surface profile depth of 40 to 100 μm. To prevent the steel from contamination, which may be introduced during the blast cleaning process, use condensate traps to remove potential contaminants (e.g., oil and water) from the compressed air that powers the blast-cleaning system. After blasting, remove residual dust either by vacuuming or by blowing down the steel with clean compressed air.

Coating application [Kehr et al., 2012]. With 100% solid materials, three methods of application exist (i.e., hand (brush, roller or squeegee), manual spray, and automated spray). Each application technique has its advantages and disadvantages. Hand application depends on the skill of the operator with both premixing and application. The process is slower and more labor-intensive, bringing operators into closer contact with the coating material and increasing the risk of exposure if proper protective/respiratory equipment are not used. Hand application is effective for small coating projects. For coatings without solvent, manual spray application requires the use of plural-component spray systems. There is less handling of the coating material, which reduces the risk of skin contact. However, spray operations require respiratory protection for workers. Manual spray is generally faster than hand application. Spray-grade coatings are designed to react faster than hand-applied ones, reducing the probability of contamination from flying debris or insects. The process is well-suited to large or irregularly shaped objects. Automatic spray application requires proportioning equipment similar to that used in a manual spray operation. Automated spray is usually suitable for uniformly mixed and shaped objects, such as girth welds for pipeline coating rehabilitation. It provides greater control over coating thickness and uniformity, thereby saving the coating material. This approach requires more set-up time, but can be much faster than either hand- or manual-spray application.

Coating curing. The time required for a coating to gel and cure depends on the specific coating material, the temperature of the steel structure, and atmospheric conditions. If the steel or ambient temperature is below the cure range of the specific coating, reheating the steel structure should be considered. Without specific instructions from the coating or material manufacturer, a good starting point to ensure cure is to preheat the structure to about 149°F (65°C) if the ambient temperature is between 14°F and 50°F (-10°C and 10°C). If the ambient temperature is below 14°F (-10°C), preheat the structure to about 194°F (90°C). If the steel part's temperature is above 194°F (90°C), care must be taken when applying coating to prevent volatilization of certain coating components. One way to avoid this problem is to first apply a thin layer of the coating (250 μm or less) and allow it to gel before applying the remainder of the coating to the specified thickness [Kehr et al., 2012].

3.1.4. Polyethylene Coatings

3.1.4.1. A Brief Look at the History of Polyethylene

The first PE synthesis practical for industry was discovered in England in 1933 [Willbourn, 1983]. Two researchers, Fawcett and Gibson, started an ethylene reaction with benzaldehyde at 338°F (170°C) and 1,900 atm. After several days, the pressure had decreased due to a leak, and all the benzaldehyde had blown out of the reactor into the thermostat oil. It was further observed that the tip of a steel U-tube was coated with a waxy material, which has been accepted as the first recorded observation of PE formation. PE emerged shortly afterward as a commercial product, and rapidly found many uses. However, it was soft and began to melt at low temperatures. The reason is that an ideal PE molecular chain would be a long chain of carbon atoms, as shown at the top of Figure 3-4. Sometimes, under high-pressure polymerization conditions, ethylene molecules do not add on in a regular fashion during intermolecular chain-transfer reactions. Instead, short branches in 2-5 carbon atoms are added in the polymer chain, as shown at the bottom of Figure 3-4.

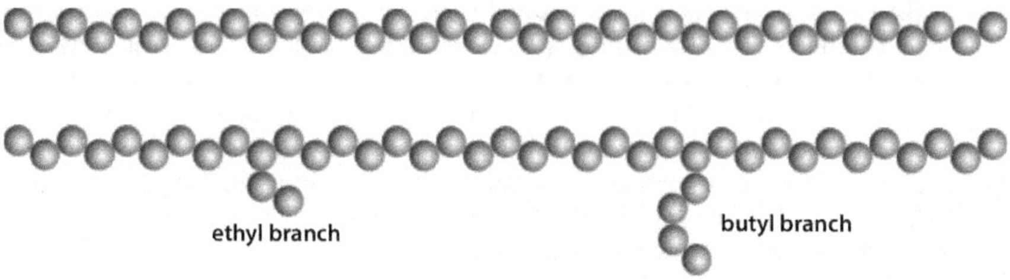

Figure 3-4. (a) Ideal polyethylene chain, and (b) short-branched polyethylene chain [Trossarelli and Brunella, 2003].

The first solid ethylene polymer, which is also the first and simplest member of the vinyl polymer family (i.e., high-pressure polyethylene), was discovered as an unexpected result of later research. Originally, the old and new types of polymers from ethylene were referred to as high-pressure or low-pressure PE according to the polymerization processes. Now, the old type of PE (the one produced through the high-pressure process) is called low-density polyethylene (LDPE). Polymers of ethylene produced at low pressures and relatively low temperatures with the aid of catalysts (based on titanium halides and aluminum alkyls), also called Ziegler catalyst, are substantially different from the LDPE. They possess a higher crystallinity and a much higher density than LDPE, and are referred to as high-density polyethylene (HDPE). When a PE contains a small number of side branches and generates a product with a moderate density, it is called medium-density polyethylene (MDPE).

Around the 1980s, with advances in catalyst technology, a linear PE was generated with a relatively low density between the MDPEs and the LDPEs. These materials were called linear low-density polyethylene (LLDPE). Since PE is a major insulator for electric cables, a new form has been developed where polymer molecules are lightly cross-linked, to prevent them from turning into liquid if a cable overheats. This kind of polyethylene is known as cross-linked linear polyethylene [Trossarelli and Brunella, 2003].

In the presence of good catalysts, it is possible to control PE's molecular weight. Commercially available PE ranges from medium molecular weights to ultra-high molecular weights, and manufacturers can now prepare grades for specific applications. For example, ultra-high molecular weight polyethylene became the major material used in artificial hip and knee joints [Trossarelli and Brunella, 2003].

Over six decades ago, PE-type products were first used as pipeline coatings. The first extruded PE was introduced in 1956 as cross-headed extruded polyethylene over an asphalt mastic adhesive [Keller, 1956]. In 1972, the side-extrusion method was introduced to accommodate an increased demand in large-size pipe coating [Johnson and Chitkara, 1973]. PE extrusions are nontoxic and have less environmental impact. Use of these systems continues to grow because of easy handling, moisture resistance, etc.

There are four types of polyethylene pipeline coatings: PE tapes, dual-layer, three-layer, and multi-component coatings. All types of coatings are extensively used in the pipeline industry, with significant variations in coating properties, effectiveness, and cost. The CSA standard Z245.21-10 provides minimum requirements for dual-layer, three-layer, and multi-component PE coatings in Canada, and breaks PE further into four additional systems (i.e., adhesive and PE outer sheath; adhesive and PE outer sheath with more stringent peel-adhesion requirements; liquid or powdered epoxy primer with a powdered copolymer adhesive and a polyethylene outer sheath; and powdered epoxy primer, a powdered copolymer adhesive with a powdered polyethylene outer layer). [Canadian Standards Association, 2010]

3.1.4.2. Properties of Polyethylene

The unit element for forming PE is ethylene (i.e., $CH_2{=}CH_2$) which is the simplest olefin. Ethylene is a stable molecule that polymerizes only upon contact with catalysts containing mainly titanium(III) chloride, the so-called Ziegler-Natta catalysts, by:

$$n(CH_2 = CH_2) \xrightarrow{\text{Catalyst}} (CH_2 - CH_2)_n$$

The product is known as polyethylene. Figure 3-5 shows the basic structure of a PE molecule [Peacock, 2000].

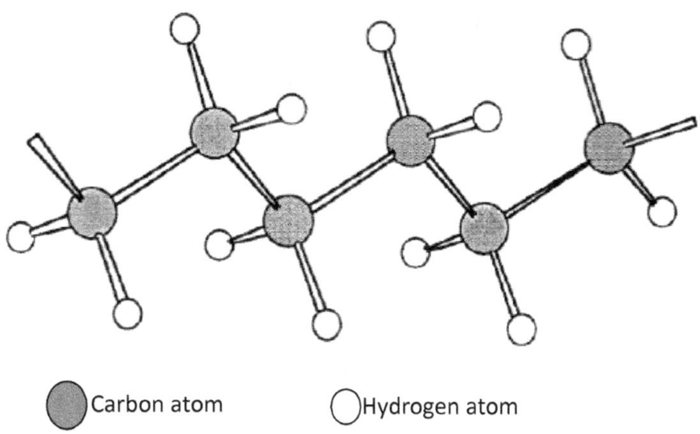

Figure 3-5. Structure of polyethylene molecule [Peacock, 2000].

Polyethylene is typically a white powdery and granular solid under ambient conditions. PE is a semi-crystalline material with excellent chemical resistance, wear resistance, and a wide range of properties determined by their structures and molecular weights. They are resistant to water, acids, alkalis, and most solvents. They have a higher impact strength, but lower working temperatures and tensile strength. In most cases, selected PE additives improve its stability and properties. For example, yellow pigments are added to some PE pellets for UV protection and other purposes for pipeline coating, and the coating is characterized by its color and is widely known as "Yellow Jacket" [Shaw Pipe, 2010].

Based on the polymerization process, PE is commonly named with the resin density and molecular weight, such as LDPE, MDPE, HDPE and UHMWPE (ultra-high molecular weight polyethylene), etc. The densities of LDPE, MDPE and HDPE are 0.880-0.915 g/cm^3, 0.926-0.940 g/cm^3, and greater or equal to 0.941g/cm^3, respectively. Generally, with the increase in density, PE has a low degree of branching, and therefore strong intermolecular forces and tensile strength. The UHMWPE has a molecular weight numbering in millions, usually between 3.1 and 5.67 million. The high molecular weight makes it a very tough material.

The adhesion of PE to smooth metal surfaces is very weak. Several practical and economical methods have been developed to modify the PE surface for improved adhesion. These include mechanical and wet chemical treatments and gas-phase processes, such as particle-beam bombardment.

The PE is nearly impermeable to water and aqueous solutions. Water vapor permeability is 1.0 g/m^{2}day for HDPE, and 5.0 g/m^{2}day for LDPE. Relative humidity does not affect the material's water permeability. For example, as temperatures change in hot, damp atmospheres, the amount of surface adsorption of water is very slight. Even the vapor permeation through PE is very low.

In terms of its physical flexibility, PE is extremely sensitive to changes in temperature, but it can be used at low temperatures of 0.4°F (-18°C) or less without risk of brittle failure. The different thermal coefficients of substrate steel and polymeric coatings cause internal stresses and stress concentration, resulting in PE coating failure.

Furthermore, PE is an excellent electrical insulator, showing a very high resistivity and low dielectric constant, all unaffected by temperature and humidity over the usual range of service conditions.

3.1.4.3. Polyethylene Tape

Polyethylene (PE) tape was first introduced into the pipeline industry in the early 1950s with the hope to provide long-lasting corrosion protection. Prior to that, cold-applied single- or multi-layer tape systems were used primarily for the protection of steel and ductile iron pipes in the water sector. Tape was first applied in the field with a technique known as Over-the-Ditch, which was later extended to in-plant application. Tape coatings are easy to apply with a portable plant setup near the pipeline route. Today, the PE tape market has undergone further growth to protect pipelines that transmit and distribute potable water. Pipe diameters typically range from 24 to 144 inches. PE tapes are used almost exclusively for protecting pipelines and tubular structures from corrosion, providing a barrier resistant to the service environment.

PE tape consists of a PE layer backed with adhesive material, usually a blend of butyl rubber and synthetic resins in a tape form. The outer layer forms a bond with the adhesive layer and is the mechanical layer, made of HDPE or polypropylene. The outer layers are usually resistant to sunlight or UV light for above-ground storage. The PE tape system relies on tension to provide a clamping effect, blocking the ingress of water and oxygen. For spiral and longitudinal weld areas on a pipeline, a strip of adhesive is often used prior to tape-wrapping to ensure the step-down of the tape from the weld beads to the pipe surface, preventing the occurrence of a "bridging effect" between adjacent welds [Romano et al., 2012].

There are two types of PE tape (i.e., laminated and co-extruded tapes). Laminated tapes are manufactured by applying an adhesive to one side of a PE film, which is usually of medium density and provides optimum strength and flexibility. The tape has a thickness varying from 375 to 1250 μm, where the adhesive portion is 125 ~ 500 μm in thickness. Co-extruded tapes are manufactured from laminated tapes. Three separate feeders carrying melted or heated tie, resin, and adhesives send raw materials to an extrusion die. The exiting products are homogeneous tape rather than laminated tape.

Hot-applied tape systems consist of a fused multi-layer coating made from a thermally activated primer layer, a thermoplastic elastomer adhesive layer, and a thermoplastic outer layer. The primer is a solvated thermoplastic elastomer that chemically bonds to the adhesive layer. The primer contains additives designed to improve the system's anti-corrosion properties. The adhesive layer fuses to the outer PE layer. The outer layer is typically a HDPE. This approach is to form an integrated system that bonds to the pipe steel and resists mechanical damage. The process, as shown in Figure 3-6, involves the application of heat to achieve the desired performance.

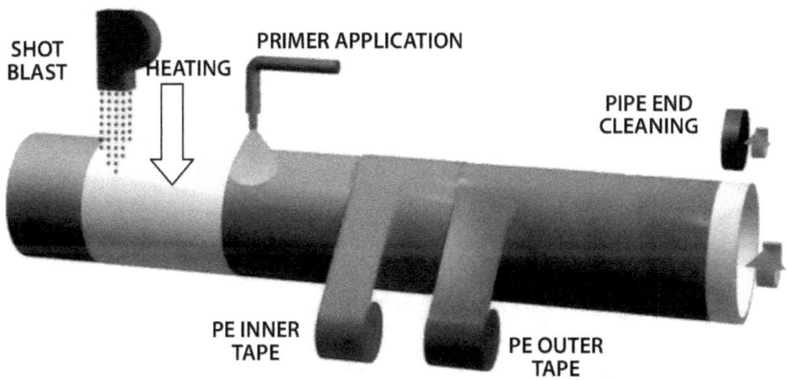

Figure 3-6. Heat-applied PE tape coating system.

The PE tape can either be applied manually for small-diameter pipelines or mechanically for larger-diameter pipelines. The coating can be used in a wide variety of sizes, thicknesses, and compositions, making it a versatile corrosion-resistant coating. In the past decades, improvements were made to enhance PE tape adhesion properties. PE tape is now primarily used on small-diameter pipelines for areas of field repair.

3.1.4.4. Dual-layer Polyethylene Coatings

A significant advantage of multilayer technology is that unique characteristics can be developed by selecting different coating layers with specific properties. Each layer can be designed and incorporated into the coating system, introducing specific characteristics that combine to significantly improve performance exceeding that of any single coating. Dual-layer PE coatings are referred to as "Yellow Jacket" and consist of a continuous outer sheath of HDPE extruded over a rubber-modified asphalt-based adhesive. Yellow Jacket provides external protection for oil/gas and water pipelines.

Dual-layer PE coatings can be applied on pipelines with diameters up to the nominal pipe size (NPS) of 12 inches and the operating temperatures of 60°C. The coatings are ideal for oil/gas distribution systems and small-diameter transmission pipelines. Figure 3-7 shows the plant application process of Yellow Jacket on a steel pipe [BrederoShaw 2].

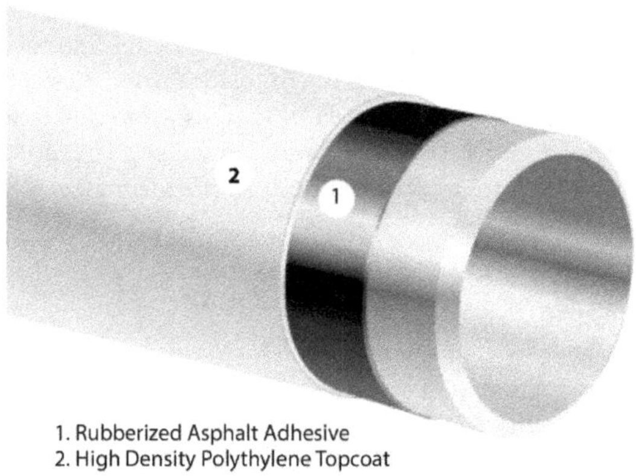

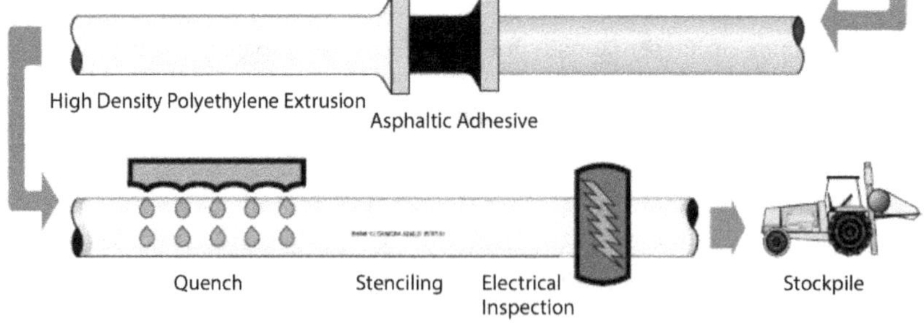

Figure 3-7. Yellow Jacket and the plant application process [BrederoShaw 2].

3.1.4.5. Three-layer Polyethylene Coatings

Three-layer PE coatings consist of an FBE primer, a copolymer adhesive layer, and either an extruded or wrapped outer PE layer. The FBE primer layer provides superior adhesion to a steel substrate. Generally, the wrapped outer PE layer is incomparable to the extruded outer layer in terms of coating performance.

Three-layer PE coatings with an extruded outer layer are commonly known as "YJ2K," and can be applied on pipelines up to NPS 12 inches with operating temperatures between -40°F and 185°F (-40°C and 85°C). Three-layer PE coatings with a wrapped outer layer are called 3LPE and can be applied to pipelines up to NPS 48 inches. Figure 3-8 shows the application process of 3LPE in plant [BrederoShaw 3].

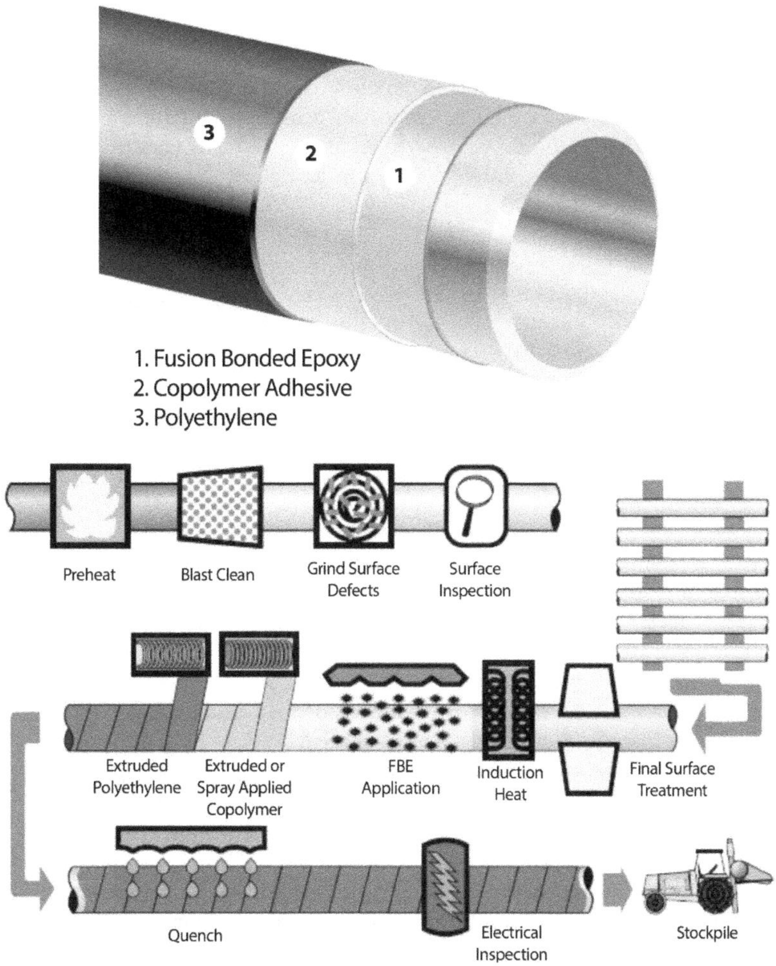

Figure 3-8. Application process of 3LPE in plant [BrederoShaw 3].

3.1.4.6. Multi-component Polyethylene Coatings

Multi-component PE coatings consist of a single layer of multiple components containing FBE, adhesive, and PE. All three components are processed and applied in powder form to create a continuous coating. These coatings are commonly known as high-performance composite coatings [Singh and Williamson, 1999]. The single-layer structure of HPCC provides superior properties over almost all other PE-type coatings, but it is also the most expensive coating. The HPCC system offers an improved resistance to moisture ingression in hot/wet environments and is coupled with the superior field-handling characteristics (including significant impact resistance) of FBE primer (i.e., a strong adhesion to the steel and outer PE).

3.1.5. Fusion-bonded Epoxy

Fusion-bonded epoxy has been the preferred pipeline coating choice (especially in North America) since the 1960s. FBE provides excellent corrosion protection for pipelines transporting water, natural gas, crude oil, diluted bitumen, and petrochemical products. Currently, FBE is used to coat the interior and exterior of pipes and field-weld joints as well as the interior of tubing and fittings [Dickerson, 2001].

FBE coatings are one-part, 100% solids, thermosetting materials in powder form that bond to steel surfaces through a heat-generated chemical reaction [Vincent, 2001]. With the exception of welded field joints, the epoxy coating is plant-applied to preheated pipes (approximately 450°F/232°C), special sections, connections, and fittings using fluid-bed, air spray, or electrostatic spray methods. Application temperatures for FBE should not exceed 525°F (275°C). Formulations of FBE consist of epoxy resins, hardeners, pigments, flow-control additives, and stabilizers to provide ease of application and performance. Due to its excellent adhesion, chemical resistance, and ease of use, FBE has been a protective coating on pipelines over several hundred thousand kilometers around the world [Kehr and Enos, 2000].

FBE can be used as a stand-alone product in a single layer or as a primer in dual-layer FBE and three-layer polyolefin systems. When applied as a single layer, the FBE requires 300 μm minimum thickness. Due to reduced bendability with increased coating thickness, the maximum thickness for single layer FBE is up to 500 μm. FBE can be applied up to 750 μm, but these pipes are not field-bent.

A primary advantage of the FBE coatings is that they do not hide apparent surface defects. Thus, a steel surface can be inspected even after being coated. The number of holidays that occur is a function of the steel substrate surface condition and the coating thickness. Increasing the thickness usually minimizes this problem. An excellent resistance to electrically induced disbondment of FBE coatings has resulted in their frequent uses as pipeline coatings.

The FBE formulations used today are substantially different from their predecessors of even a few years ago. Generally, environmental conditions where coated pipelines are operated and the expectations for coating performance have changed with time, often increasing in severity. FBE coatings have continuously evolved to meet these new challenges. Most notably, adhesion and resistance to handling damage have improved remarkably without compromising other properties of the coating

[Kehr and Enos, 2000]. It was commented [Dickerson, 2001] that "FBE coatings have revolutionized the pipe coating industry."

3.1.5.1. A Brief History of FBE Pipeline Coating

FBE was first introduced by the 3M Company for pipeline corrosion protection in the late 1950s. The first product was based on solid EPON resin, cured with dihydrazide, and showed excellent corrosion resistance but was somewhat brittle. In 1965, 3M introduced an improved product based on EPON resin, cured by methylenedianiline, and improved coating flexibility. At the same time, Shell's Union Technical Service Laboratory generated interests in EPON resins for FBE coatings by developing three promising systems, all based on Shell's EPON resin #1004, with different curing agents (i.e., benzopherone tetracarboxylic dianhydride [BTD], trimellitic anhydride (TMA), and dicyandiamide [dicy]). The BTD system facilitated the fastest curing with the best overall performance of any anhydride system, but it was deemed too expensive to compete with existing systems. The dicy system showed good performance, but was associated with slow curing and did not have enough heat capacity to cure a coating without post-baking. The TMA system provided excellent chemical and physical properties and could be formulated at a competitive price.

The field testing of FBE coatings to determine their performance under operating conditions (both underground and aboveground) was started in the 1960s. The first test was conducted on 56,000 feet of 6-inch pipeline installed in Puerto Rico in the winter of 1965–1966. The pipeline was coated on the inside and outside in the plant with EPON resin powder. An EPON #1004/TMA powder was used on the exterior, and an EPON #1004/BTDA powder on the interior. Epoxy-powder coating was applied to the hot pipe by electrostatic spray. The heating was sufficient to melt and cure the coating in less than one minute. Pipes were then water-cooled and inspected for imperfections, which were repaired on the spot.

Another test installation of the EPON #1004-TMA coating was made in Bermuda in September 1966 on 7,000 feet of 6-inch aluminum pipe that carried aviation fuel. The coating was applied on the exterior only.

While 3M's epoxy powder coating and application process was suitable for coating large-diameter pipelines, weld joints were the weakest points to be coated. In 1975, Commercial Resins Co. started working on a prototype unit to coat internal girth welds with FBE. In 1977, the 24-inch and 30-inch Tennessee Gas pipelines laid in the Gulf of Mexico were the first large-diameter pipelines to have girth welds coated internally and externally with FBE [Dickerson, 2001].

In 1999, there were three major U.S. suppliers of FBE powder coatings (i.e., 3M, Dupont Powder Coatings, and Lilly Industries). About 10 companies applied FBE coatings to oil and gas pipeline in the U.S., and there were about the same number outside the U.S. with the number of applicators growing. A few other companies make and apply FBE to their own pipes, mostly internal coating for downhole oil-well pipes.

FBE coatings have been in use for almost a half-century. The coating formulation is continually evolving, driven by new operating environments and conditions where pipelines operate, as well as performance expectations.

3.1.5.2. Properties of FBE Coatings

Like other powder coatings, FBE contains essential components of resin, hardener or curing agent, fillers and extenders, and color pigments. The resin and hardener components are known as the binder. In FBE coatings, the resin is an epoxy-type resin. Epoxy structure contains a three-element cyclic ring (i.e., one oxygen atom connected to two carbon atoms) in the resin molecule. Most commonly used FBE resins are derivatives of bisphenol A, as shown in Figure 3-9 [Wikipedia]. Resins are also available in various molecular lengths that provide unique properties to the final coating product.

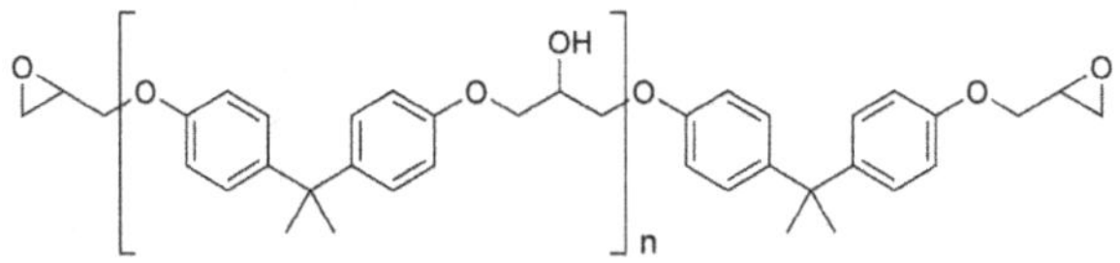

Figure 3-9. Structure of unmodified Bisphenol A type epoxy prepolymer, where n denotes the number of polymerized subunits and is in the range from 0 to about 25 [Wikipedia]

The curing agents contained in FBE react with either the epoxy ring or the hydroxyl groups. The selected curing agent determines the nature of the final FBE product (such as the cross-linking density of the coating and its chemical resistance, brittleness, flexibility, etc.). The ratio of the amount of epoxy resins to that of curing agents in a coating formulation is determined by their relative equivalent weights.

In addition to these two major components, FBE coatings include fillers, pigments, extenders, and various additives to provide desired properties. These components control characteristics including permeability, hardness, color, thickness, gouge resistance, and others. These components are normally dry solids, even though small quantities of liquid additives may be used in some FBE formulations.

FBE is now the first choice of pipeline coatings used in North America. The coating possesses many properties, making it an excellent coating for pipelines. These include strong adhesion to steel substrate, good flexibility, high chemical resistance, low oxygen permeability, compatibility with CP, and a wide usable temperature range (from -40°C to 105°C). Moreover, there has been no reported case of SCC on FBE-coated pipelines over the last 40 years [Been, 2011].

The major concern with the use of single-layer FBE coating is transportation and handling damage. To improve its impact and abrasion resistance, the coating's thickness can be increased. However, the coating's flexibility decreases with its thickness. Other major drawbacks of FBE coatings include high moisture absorption (although this has proven to be an advantage when used with CP), difficulty in field joints, and the requirement of high thickness to eliminate weld tenting.

3.1.5.3. Application of FBE Coatings

FBE, as the most commonly used coating for new, large-diameter oil/gas pipelines, is a heat-activated and chemically cured coating system. Prior to coating application, the pipe's steel surface must achieve the white, blasted clean standard state [NACE/SSPC, 2010] to provide an appropriate anchor pattern to enhance the coating adhesion. Figure 3-10 shows the FBE application in-plant process [BrederoShaw 4]. Some systems may require a primer, and some require post-heating for complete cure. The steel pipe is pre-heated up to 150°F (67°C), to remove moisture. The pipe is then shot and grit-blasted for a clean surface with a proper anchor-pattern profile and improved adhesion properties. The pipe's surface condition must reach the NACE/SSPC standard [NACE/SSPC, 2010]. Further grinding may be required to remove surface imperfections, and a vacuum cleaning or blowing with dry, clean air removes dust from the pipe surface. The pipe is then heated up to 438°F–475°F (225°C–246°C), and the FBE is applied by electrostatic spray process, which melts and flows over the steel surface. The deposited, melted powder generates a smooth, tough, and solvent-free finish. The coating can cure rapidly. The pipe is then cooled and becomes ready for handling, inspection, and installation. The FBE is typically applied to the steel pipe in thicknesses between 300 μm and 500 μm. For thicknesses greater than 400 μm, the coating is restricted not to bend. For dual-layer coating systems, a second set of guns applies coating in-line to the top layer immediately after coating the primary layer.

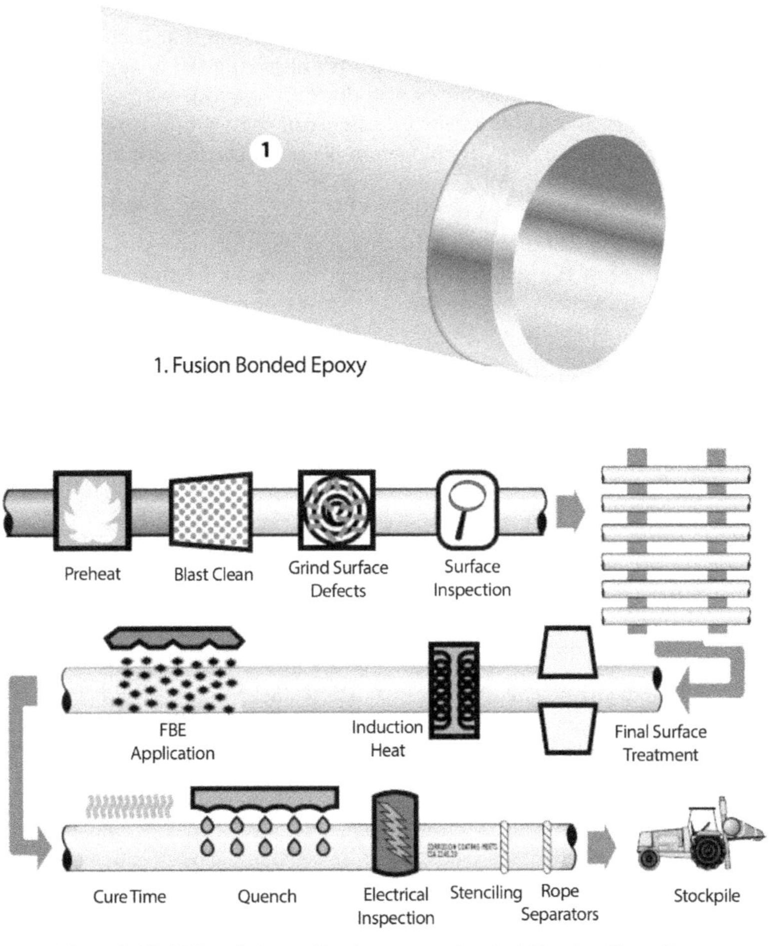

Figure 3-10. FBE and the application process in plant [BrederoShaw 4].

Liquid-epoxy coatings are sometimes applied over the FBE to protect pipelines against abrasion and mechanical damage during construction.

3.1.5.4. Single-layer FBE Coatings

FBE coatings are commonly used as monolithic corrosion-protection coatings. It has already been determined [Goldie, 2010] that the good barrier performance of FBE is due to its hydrophilic nature and ability to maintain a high glass-transition temperature (T_g). The key to maintaining adhesion results from good surface preparation, usually based on the chemical pre-treatment of blasted steel with chromate- or phosphate-based products. Strong adhesion is one of the FBE coating properties that has improved significantly over the years. Another feature of FBE is its capacity for adhesion recovery, which is often seen in bell hole evaluations.

In addition to its favorite properties, FBE is a permeable coating that is conductive to CP current. Its expected performance is in good adhesion with no blistering. However, when the coating loses adhesion or generates blisters, the steel beneath it does not corrode with adequate CP. Steel discoloration is common, but the pH of the electrolyte beneath the coating is normally basic (high pH). It was suggested [Neal, 1998] that steel discoloration is due to the development of a magnetite layer, which is an important component in pipeline protection.

The application process of a standalone FBE is quite straightforward as previously described. The technique is prone for use in the field with portable heating and application equipment. Field-application of FBE on girth welds provides the same level of performance quality as plant-applied coatings. Consequently, the pipeline can be protected with the same coating from end to end [Kehr and Enos, 2000]. However, standalone FBE has less damage tolerance, which can be addressed with increased thickness of the coating. Single-layer FBE is thin compared to other coatings, so damage during transportation and pipeline installation could be caused, resulting in higher repair and CP costs. When properly specified, inspected, applied, and handled during the various stages of application, storage, transportation, and construction, this damage is minimized. Unlike the thicker, heavier coatings, damaged FBE can easily be located and repaired. When there is an issue with handling, storage, and construction damage, either a dual-layer FBE system or a three-layer polyolefin with FBE primer should be considered.

In summary, the single-layer FBE has been a capable pipeline coating since its introduction in 1960, and is now the most commonly used pipeline coating in North America, with a large following in the rest of the world. It not only has performance characteristics important to the application and construction processes, but it also has proven performance in underground and undersea services. It has proven effective for pipelines, girth welds, fittings, and bends. When used at a greater thickness, it has worked effectively with impinged concrete and directional-bore installation.

3.1.5.5. Dual-layer FBE Coatings

FBE coatings can be applied as dual-layer powder coatings, increasing their resistance to friction and abrasion. Directional drilling for road and river crossings has become common in pipeline installation and repair. During installation, the pipe is pulled through a hole supported by drilling mud, where coating abrasion and scratching can occur. The application of dual-layer FBE coatings can mitigate damage to the base FBE during installation [NACE, 2011].

In dual-layer FBE coatings there is a second layer of FBE on top of the base coating. During application, the second layer (or topcoat) is applied directly after the base layer, when the base layer application still has a chemical bond with the steel surface. Typical thickness of the dual-layer coating is up to 1,000 μm. In 2002, the first major pipeline was coated with a dual-layer FBE coating from end-to-end, including girth-welds [Pratt et al., 2011].

Development of two or more layers of FBE coatings improves the coating's performance and protective capability. The base layer provides a good initial adhesion that is maintained after exposure to environmental conditions. The top layer can provide many more features, including:

- improved resistance against mechanical damage (e.g., impact, abrasion, friction and gouge, generated during transportation and construction)
- enhanced high-temperature performance (i.e., the combination of an adhesion-enhanced basecoat with a thick topcoat layer ensures high-temperature performance up to 230°F (110°C), with improved cathodic disbondment resistance)
- improved UV resistance (when exposed to humidity and sunlight (or UV radiation), chalking occurs with FBE coatings on pipelines, but powders in the top layer have good chalking resistance)

3.1.5.6. FBE as Primer for Three-layer Systems

The use of three-layer polyolefin coatings for pipeline protection began in Europe around 1980. Multi-layer coatings overcame the weakness of single-layer coatings through the advantageous characteristics of material combinations. Generally, development of multi-layer coatings is based on two principles: Coatings optimized with a multi-layer design that combines the favorable properties of individual coating materials, and the functional separation of corrosion protection and resistance to mechanical damage [Singh and Cox, 2000].

The basic composition of a three-layer polyolefin coating includes FBE as a base layer (primer), an intermediate adhesive tie layer, and a polyolefin (either PE or PP) top layer. PE is commonly used for temperatures up to 175°F (80°C). For temperatures between 175°F–230°F (80°C–110°C), PP is used. The PP is also usually used when better mechanical properties are required for handling. FBE primer is used for its excellent adhesion to steel substrate and cathodic disbondment resistance. Polyolefin is a non-polar molecule and has low moisture-absorption and good electrical-insulation properties. The tie layer or adhesive layer is typically polyolefin modified with polar end groups grafted onto carbon backbone and is compatible with the outer layer of PE or PP. Interlayer adhesion is achieved with a chemically modified polyolefin material with polar end groups, which can form linkages between the non-polar polyolefin and the polar FBE [Singh and Cox, 2000].

Three-layer coating systems are applied in-plant, following steps similar to the FBE application. An FBE layer is applied by electrostatic spray. Induction or gas heating brings the pipe to required surface temperatures for proper application of the powder. An adhesive layer is then applied by side extrusion or spray, followed by the polyolefin outer layer. The FBE primer base coat can be 100–200 μm in thickness. The thicknesses of the extruded copolymer adhesive and the extruded polyolefin outer layer depend on operating temperature, pipe size, and the polyolefin used [NACE, 2011]. The typical outer-layer thickness ranges from 500 μm to several millimeters. The thickness can be up to 3 mm for three-layer coatings with an LDPE outer layer and used in severe transportation and construction scenarios. However, a lower thickness is recommended, especially for MDPE and HDPE outer layers that have better impact resistance [Singh and Cox, 2000].

Although a top-layer coat of polyolefin has large mechanical strength and low moisture permeability, making it suitable for impact and abrasion resistance, it is not possible to pass CP current through it due to its insulating effect. Thus, if the coating is degraded (disbonded) and corrosion occurs beneath it, CP effectiveness is significantly reduced. Moreover, if the cohesive bonding between the layers fails and electrolyte gets trapped between them, the coating system's integrity is compromised [Kehr and Enos, 2000].

3.1.5.7. Further Development of FBE Coatings

The development of new FBE coating products has been driven by the heavy dependence on coating by the maintenance of pipeline integrity. The newly developed FBE can be applied at temperatures as low as 356°F (180°C). However, for FBE coatings to achieve optimum performance, they need to be applied at temperatures in excess of 446°F (230°C) for single-layer systems and 392°F (200°C) for three-layer systems. This indicates the direction for development of new FBE formulations to balance temperature requirement and operating performance.

High-strength steel pipelines have started to develop in recent years. However, most of the high grades of steel cannot withstand pre-heat temperatures above 392°F (200°C). The temperature limit constrains coating use, particularly with the use of FBE coatings. New FBE products will provide a solution for protecting pipelines made of high-strength steels.

Furthermore, development of a mid-T_g coating suitable for applications of high temperature pipelines 248–302°F (120–150°C) with good flexibility and adhesion has been an interest. As exploration and extraction of oil from fields at increasing depths continues, pipelines need to carry fluids at temperatures in the 248–302°F (120–150°C) range. It is widely accepted [Goldie, 2010] that the T_g of a coating should be at least 18°F (10°C) above the operating temperature of the pipeline, but few FBE coatings are commercially available with T_g in the 266–320°F (130–160°C) range. While a few coatings can withstand these high temperatures, they have limited flexibility, particularly at low temperatures. The low flexibility limits the range of environmental conditions in which coated pipelines can be used. Thus, the development of a range of mid-T_g FBE powder coatings (i.e., T_g in the 266–320°F (130–150°C) range) that possess excellent adhesion and mechanical properties at temperatures even down to-76°F (-60°C) has been the new technology in FBE coating innovation.

3.1.6. High-performance Composite Coating

With the increasing energy demands, there have been growing pipeline activities in remote regions such as Arctic and sub-Arctic areas in North America. The development of innovative coating technology can address environmental and operational challenges including cold climates, permafrost and semi-permafrost, and difficulty with maintenance. To ensure pipelines' long-term integrity in the northern area (where severe conditions affect pipeline construction, design, and operation and influence pipeline coating selection and performance), an innovative, multi-component coating known as high-performance composite coating was developed [Singh and Williamson, 1999; Singh et al., 2005].

The HPCC is a powder-coated, multi-component, single-layer coating consisting of an FBE primer as the base coat, an MDPE outer layer as the top coat, and a tie layer containing a chemically modified PE adhesive [Singh and Cox, 2000; Howell and Cheng, 2007]. All components in HPCC are applied in powdered form using an electrostatic powder-coating technique. The tie layer takes the form of a physical-interlocking mechanism between the PE topcoat and the FBE base coat, generating a well-defined, smooth-transition layer. The single-layer HPCC reduces the risk of interlayer delamination (i.e., cohesive failure of coatings, a key failure mechanism usually found with some 3LPE). Moreover, as the coating is applied in powder form, its thicknesses can be easily customized to meet certain criteria or coating requirements.

3.1.6.1. Structure and Composition of HPCC

The HPCC coating is a single-layer, multi-component coating consisting of an FBE primer, an MDPE outer layer and a tie layer containing a chemically modified PE adhesive, with all applied as powders. Figure 3-11 shows a schematic diagram of the cross-section of an HPCC coating system with a total standard thickness of 750 μm [Singh and Cox, 2000]. The minimum thickness recommended for the FBE component layer is 175 μm, but other thicknesses in the range of 100–400 μm have been used [Singh et al., 2005]. The adhesive interlayer for bonding the FBE primer and PE topcoat together has a minimum standard thickness of 150 μm. It was reported [Singh et al., 2005] that thicknesses between 50–72 μm have been used, since only a small amount of adhesive material is sufficient to form a good chemical bond and blend all three components to form a composite material. The PE outer layer has a minimum thickness of 500 μm. The standard thickness of the PE topcoat can be increased several more millimeters to withstand environmental conditions or to meet several coating requirements (such as mechanical protection, chemical resistance, and weather resistance). For heavy-duty applications, the thickness of the PE coat can be increased to 1,250 μm for a total standard thickness of 1,500 μm. This is often referred to as an HPCC heavy coating [Guan, 2011]. Generally, the corrosion protection of an HPCC system is linked to its total thickness. However, increasing the coating thickness can increase the total cost. Hence, a balance between the coating cost and the thickness should be made to achieve a low cost but still maintain high performance and coating integrity.

Coating Cross-Section	Material	Standard Thickness	
		µm	(mils)
	Polyethylene topcoat	500	(20)
	FBE/Adhesive Interlayer	125	(5)
	FBE	125	(5)
	TOTAL	**750**	**(30)**

Figure 3-11. Schematic diagram of the cross-section of a HPCC coating system with a total standard thickness of 750 µm [Singh and Cox, 2000].

Figure 3-12 shows microscopic images (400x magnification) of the FBE and PE components and their interface in an HPCC coating [Howell and Cheng, 2007]. The PE layer has a more compact structure with fewer pores or holes compared to the FBE part. Moreover, there is no defect (e.g., pore, bubble, or hole) found at the interface between the FBE and PE parts. Since the adhesive interlayer is thin and has a similar color to the outer PE part, it is indistinguishable.

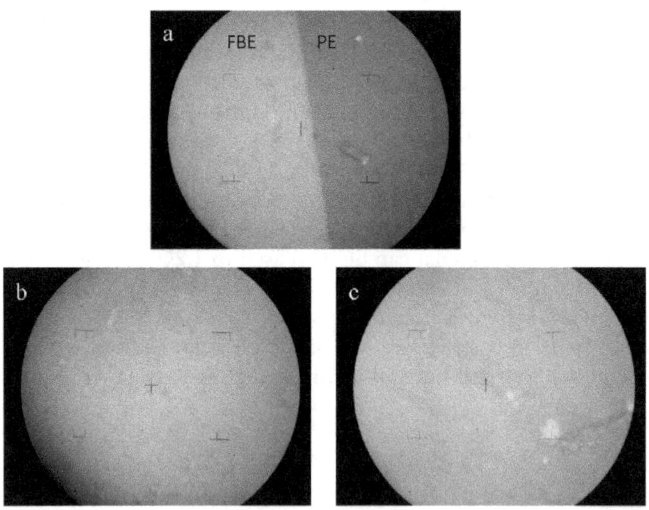

Figure 3-12. Optical view of the HPCC structure and morphology at 400 times of magnification (a) PE/FBE interface, (b) FBE part and (c) PE part [Howell and Cheng, 2007].

3.1.6.2. Properties of HPCC

Generally, the design of multi-component coatings is based on two principles (i.e., optimization by combining favorable properties of each component contained in the coating, and separation of functional performance of individual components). The HPCC coating systems provide numerous benefits for buried pipelines, including:

- superior adhesion due to excellent adhesion properties offered by the FBE primer
- good impact resistance offered by the PE topcoat to protect against mechanical damages due to field handling and construction
- enhanced resistance to water permeation to protect against moisture ingress
- good cathodic disbonding resistance
- high-temperature performance (up to 175°F, or 85°C) and flexibility characteristics for cold weather installation (down to -40°F, -or 40°C)
- ability to withstand field bending to a certain degree at cold temperatures without significant damages to the coating
- superior corrosion protection, and low risk of interlayer delamination
- prevention of weld "tenting" and cross-sectional weakness by providing an optimum conformity to raised welds at weld-joint locations

Field experiences show that multi-component powders such as HPCC offer a potential solution for protecting pipelines from corrosion attack and mechanical damage where severe handling and service conditions exist.

3.1.6.3. HPCC Application Processes

The application process for powder-coated HPCC coating consists of steps similar to FBE coating's process. Figure 3-13 shows the composition of HPCC and a typical coating application process [BrederoShaw 5]. It consists of steps similar to those of an FBE coating-application process. The steel pipe is preheated in a hot-water rinse at 176°F (80°C), giving the pipe a uniform initial temperature. Contaminants are also cleaned off in the heating process. The pipe is then abrasively blasted clean to remove mill scale, dirt, dust, rust, or other foreign matters, achieving a near-white metal finish. The surface of the pipe is inspected to ensure there is no mill-scale, rust, or cracks, and to verify that the pipe is suitable for coating. The pipe undergoes a final surface inspection to ensure all defects are removed and is followed by surface treatments such as acid washing and rinsing to promote coating adhesion to the pipe and improve surface roughness. Induction heating supplies heat to the steel pipe through electrical induction to facilitate a proper powder-coating temperature. After the coating cures, it is quenched by water to solidify the coating, remove heat from the pipe and coating, and promote coating adherence to the steel surface. Final steps include inspecting the coating for holidays, removal of water from the pipe interior, measurement of coating thickness, and quality conformance tests [Singh and Cox, 2000].

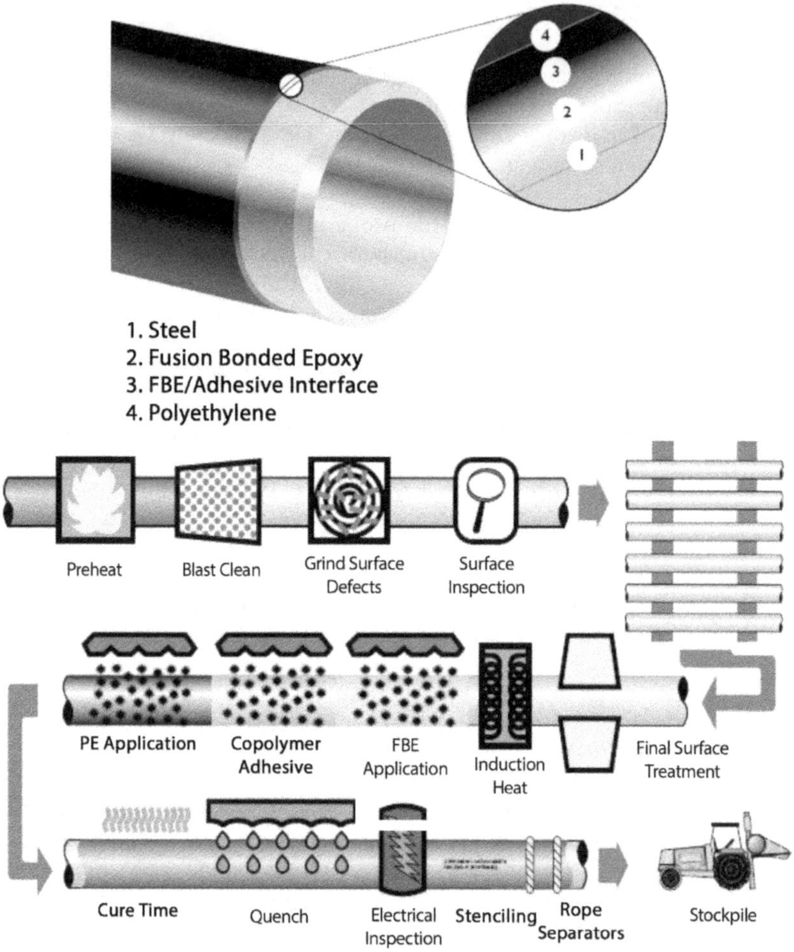

Figure 3-13. HPCC coating and the application process in plant [BrederoShaw 5].

3.1.6.4. HPCC Repair

The procedures to repair a powder-coated HPCC coating are similar to those used on conventional three-layer coatings such as 3LPE [Singh and Cox, 2000]. For minor damages to the outer PE, the area is cleaned to remove any foreign material. The PE is reheated locally, and a hot-melt stick repair is performed. Where there is damage through to the FBE primer with a diameter of the damage less than 25 mm, a two-part liquid epoxy primer is applied, followed by a hot-melt stick or patch material. If the diameter of the damage is larger than 25 mm, a two-part liquid epoxy primer is applied, followed by a hot-melt stick or heat-shrink sleeve.

3.2. Field-applied Pipeline Coatings

Applying pipeline coatings in the field requires proper selection and application techniques to ensure expected service. Since field application environments vary throughout the construction process, the selection and application of field coatings should be well prepared to deal with the constant changes. Many types and brands of coatings can be applied in the field, but not all can be applied under all environmental conditions.

Two environmental conditions that must be considered for coating application in the field are temperature and humidity.

Temperature. Temperature of the surrounding air is critical to many coatings and the rate at which they cure. In a too-cold environment, cure may not occur. If it is too hot, the cure may occur prior to proper wetting of the substrate surface. Temperature of the pipe surface is also critical for the same reasons. If the pipeline is new or out-of-service, the pipe can be heated if it is too cold for proper cure. Sometimes the area must be tented to provide adequate air and pipe temperature. In-service pipeline temperatures are usually controlled by the temperature of the product inside the pipe. This temperature is not easily changed or controlled. Thus, coating selection criterion is critical for the particular condition. Condensation is a consideration when the product (or pipe surface) is colder than the surrounding air, when dew point testing is required.

Humidity (rain and snow). Most pipeline coatings will not bond to a wet pipe. In cases with humidity problems, the coating operation must be suspended or the pipe must be tented for a controlled environment. Checking dew point and controlling temperature can help, but moisture on the pipe surface requires discontinuation of the coating application until the environment is conducive to proper application. There are very few coatings that can be applied to a wet surface. These must be researched and tested to determine if they will work in a particular situation and environment. Since CP will typically be used, the coating must be tested to determine if it will work with CP after the coating is applied to the wet surface.

3.2.1. Liquid Coating Systems

Many liquid coating systems are available for maintenance and coating repair. Rehabilitation coatings for pipelines and internal tank coatings are often liquid-applied systems, and are not generally plant-applied. Liquid coating systems include two parts of epoxy, polyurethane, epoxy/urethane, epoxy novalac, coal-tar epoxies, and coal-tar polyurethane systems. In addition to meeting performance requirements, a liquid coating must be user-friendly and able to be applied in the field under a wide variety of conditions.

Liquid epoxies have been developed to protect pipelines operating at temperatures up to 302°F (150°C). Additionally, advances in epoxy technology allow some to be applied to wet surfaces or bond to PE coating systems. Liquid coating systems are used for coating girth welds, valves, fittings, pipes, ballast tanks, water tanks, crude oil and product tanks, ships, and marine structures. Liquid coatings must be flexible enough to allow installation, have good adhesion when applied to properly prepared surfaces, be able to resist impact damage during burial, and be compatible with CP. Some liquid coatings can be applied to damp or wet surfaces, but must be properly tested and selected for these services.

As with most high-performance coating systems, surface preparation is critical to the performance of liquid-applied systems. For buried or submerged liquid coating systems, a NACE No. 1/SSPC-SP 5 "White Metal Blast Clean" is the best surface preparation to assure good coating performance. Normally, abrasive blast cleaning is used in the field to obtain the NACE No. 2/SSPC-SP 10 "Near-White Metal Blast Cleaning" surface. For larger projects, like pipeline rehabilitation or internal tank lining, high-pressure water-jetting or grit blasting may be used. During field surface preparation, the metal surface should be at least 5°F (3°C) above the dew point. If flash rusting occurs, the surface should be re-blast cleaned.

Most two-part epoxies will not cure at temperatures below 45°F (7°C), therefore, surface and air temperatures must be monitored at all times during epoxy application. Some formulations will cure below these temperatures, but these are different formulations than most two-part epoxies and should be tested to ensure proper cure and performance. When the pipe surface is heated in cold weather to accelerate the cure, the heat may start curing too quickly, which in turn may prevent the two-part epoxy from properly wetting the surface. With curing that starts too quickly, adhesion and performance may not meet the required performance level. If the air temperature is too high, the epoxy can cure in the container so much that it does not wet the surface properly.

Field application of coatings can be accomplished by brush, roller, trowel, conventional air spray, airless spray, or plural component spray equipment. Field coatings are often applied in one coat. Girth welds can be coated using brush, roller, airless spray, or plural-component spray techniques. Brush and roller coating can result in non-uniform coatings, with thin and thick areas. Airless spray and plural-component spray equipment usually results in more uniform coatings.

Coating repairs for small areas, less than 6 cm^2, can be repaired using power-tool cleaning and brush or trowel techniques. For larger areas, surfaces can be prepared with abrasive blasting followed by brush, roller, or spray coating.

Pipeline rehabilitation can require coating short or long sections, from several feet to several kilometers of pipeline. For short sections of pipeline, the pipe can be prepared through abrasive blasting and coated using airless spray or plural-component spray techniques. For longer sections of pipeline, the surface can be prepared using line-travel blasting equipment. To ensure rapid coating cure, plural-component coating techniques are often used.

3.2.2. Tape Coatings

The two basic types of tape coatings are solid-film backed and mesh backed. The solid-film backed tapes were developed first, but had major problems with disbondment (especially with soil stress) and CP shielding. The mesh-backed tapes were developed to help solve the problems with soil stress and CP shielding when there is coating disbondment.

3.2.2.1. Solid Film-backed Tapes

In the field, solid film-backed PE and mesh-backed tapes may be applied in a spiral or a cigarette wrap either by hand or by a portable wrapping tool that provides the proper tension for the tape. A cigarette wrap of tape requires the tape to be cut to a length equal to the pipe circumference plus the minimum overlap. Tensioning of the cigarette wrap helps maintain the proper overlap. Cigarette wrap is used when the pipe cannot be excavated sufficiently deep to allow for a roll of tape to pass under the pipe. With cigarette wrap, overlaps must be staggered to prevent multiple overlaps of the tape. Typically, overlaps are alternated between 10 o'clock and 2 o'clock on the pipe. All overlaps should end with the top lap in the downward position to prevent water penetration and soil stress from causing the tape to be forced off the pipe.

Spiral wrapping by hand or with a wrap machine is most often used to apply tapes. Tension on the wrap is important to keeping the tape flat on the pipe. Wrap machines help applicators apply consistent tension and overlap. As the pipe is spiral-wrapped, tensioning helps to reduce wrinkles, air pockets, and enhance adhesion. The pipe is cleaned, primed with compatible primer, and then wrapped. As with all coatings, environmental conditions must be monitored to ensure they are in the appropriate limits for the tape being applied. Typical overlaps are at least one inch, but for some applications, the overlap is 50% or more.

Frequently, tapes have release paper on the adhesive side of the tape that must be removed before the tape is applied to the pipe. This prevents the layers of tape from sticking together and keeps the amount of adhesive uniform. Tapes without release liners make for much easier product application, but they may not provide a proper seal at the overlaps, since it does not bond to the backing.

Many of the early solid film-backed tapes had backings that did not have good tensile strength, allowing the tape to stretch easily during soil stress and pipe movement. This, then, allowed electrolytes to coat the surface between the tape and pipe, causing CP shielding. Most tape manufacturers now use better compounds and solid film backings. The amount of stretch is a critical issue during coating-tape selection. Less stretch during application means less stretch during service.

3.2.2.2. Mesh-backed Tapes

Laminated mesh-backed tapes are manufactured by applying a compatible adhesive to one side of a pre-fabricated film of woven (mesh) plastic (polypropylene). Figure 3.14 shows the tape-coating lamination process. The film is designed so polypropylene provides a high tensile strength, resulting in a good resistance to soil stress. The woven nature of the backing allows the tape to be flexible enough to allow for easy installation. As with most tapes these have a release liner that must be removed before application, but can be applied without complicated equipment and requires less surface preparation than most other coatings.

Time, temperature, and tension (3 Ts of tape coating) are critical to understanding how mesh-backed tapes perform. The strength of the mesh backing allows these tapes to be applied with greater tension than the solid film-backed tapes, creating a greater hoop stress that allows the compound to fill the helix area and other voids that may exist. This tension also enhances the adhesion with time. The temperature is also critical to how easily the compound will flow and seal the pipeline from the environment.

The mesh-backed tapes are not susceptible to CP shielding like solid film-backed tapes and shrink sleeves when disbondments occur. Mesh-backed tapes have over 25 years of proven service even if disbondments (though rare) do occur. [Norsworthy, Hughes, 2007] [Norsworthy, 2004]

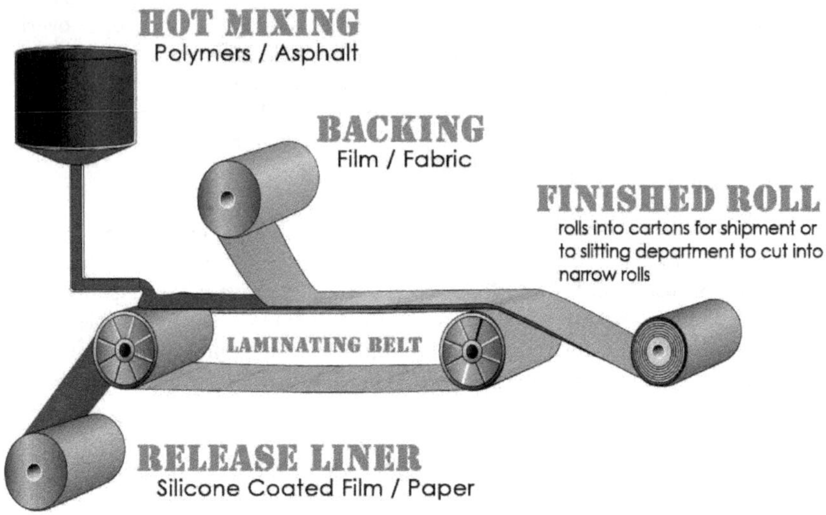

Figure 3-14. Lamination process for tape manufacturing.

Mesh-backed tapes are easily applied without complicated equipment and require less surface preparation than two-part epoxies. In recent years, the mesh-backed coating systems have become popular for girth-weld coatings on new pipelines coated with FBE, extruded polyolefin, or three-layer coatings. As with nearly all tape coatings, mesh-backed tape can be immediately backfilled after application.

Covering weld seams with a strip of tape is recommended to prevent tenting over girth or longitudinal weld seams. The other option is to use a 50% overlap and provide more compounds to fill those areas around the weld seam that might tent. When applied with proper tension, the mesh backing causes the soft adhesive compound to fill all tented or void areas left after the application. Trapped air or hydrogen from welds can escape through the soft compound and openings in the mesh backing. The tension causes the compound to reseal these areas, preventing water penetration.

Tape coatings are best applied using a hand wrapster or another wrapster type that provides the proper application tension and overlap while removing the release liner. It is more difficult to achieve proper tension during application, with an increased chance of having wrinkles or improper overlap when applied by hand. The proper width of tape should be used for the size of the pipe being coated.

3.2.2.3. Field-applied Tape Coatings

Tape coatings applied in the field can be used as mainline coating, rehabilitation coating, and joint coating. Currently, most tape coatings applied in the field are applied with squirrel-cage type machines (mechanical or by hand) or with hand-wrapping machines.

For mainline construction (rarely performed in the field in the current market), the pipe is welded into a continuous length and bent to conform to the contour of the ditch. Frequently, two side-boom tractors, which travel about 50 feet (15 m) apart along the open ditch, support the pipe and the tape application machine. The machine improves tensioning and overlap controls. The front side-boom carries a cleaning and priming machine that fits around the pipe and adjusts to the pipe diameter, while the rear side-boom carries the tape-wrapping machine that also fits around the pipe. The two side-booms lift the pipe, and as they travel forward, the pipe is cleaned, primed, and wrapped with the tape.

Alternatively, a combination cleaning/priming/wrapping machine may be used. These specially built units clean and prime the pipe with a fast-drying primer, then simultaneously wrap the tape over the primed surface. These are rarely used today.

In pipeline rehabilitation, a combination machine is used and the application procedure is similar. Because the pipe is still in-service and in its original ditch, the process includes some exceptions. Only short sections can be cleaned and wrapped. The machine travels along the pipe in the ditch, but it must frequently be taken off the pipe to bypass the necessary pipe supports.

3.2.3. Shrink Sleeves

Shrink-fit sleeves are similar to solid film-backed tapes, but made in sheet form from the same composites. A sleeve's plastic backing is chemically treated or is irradiated to cause it to expand. When the material is heated, it shrinks into a tighter and denser film as it tries to return to its normal dimension. Some of these materials are tubular and must be slipped over one end of the pipe. Most sleeves are wraparound lengths of PE backing, overlaid with a special adhesive.

The sleeves are sized to the pipe diameter, plus an allowance for overlap. They are wide enough to cover the bare field joint and extend over the coating at each cutback. The sleeve may be applied over a special adhesive primer, usually an epoxy-type primer. The wraparound sleeve is then placed around the pipe circumference with the overlap facing the ditch bottom. The sleeve is heated with a propane torch. As it heats, the sleeve shrinks to a tight fit over the joint.

The general procedure to apply sleeves include surface preparation, heating the pipe to approximately 120°F (49°C), centering the sleeve over the joint and heating it with a torch so it shrinks around the pipe diameter. The torch is kept moving to avoid burning the sleeve, and moved circumferentially from the center to the opposite end.

3.2.4. Petrolatum and Wax-coating Systems

3.2.4.1. Petrolatum and Wax Tapes

Petrolatum tape consists of a plastic fiber felt saturated with petrolatum. Wax tape is a plastic-fiber felt, saturated with a blend of microcrystalline wax. Both products can contain plasticizers and corrosion inhibitors. Petrolatum and wax tapes are typically used for corrosion protection to above- and below-ground pipelines, flanges, valves, and related surfaces. They can also be used for atmospheric and marine structures.

The tape is applied cold (no heating needed) by hand over a compatible primer. It can be applied over a damp surface. It can also be applied to wet surfaces, providing a coating system with easy application from arctic to tropical temperatures.

For field application, the pipeline surface is prepared with the removal of all loose scale, rust, or other foreign matters. High-pressure water of 21 to 48 MPa (3,000 to 7,000 psi) is suitable for the purpose. A thin uniform coat of petrolatum paste or primer is applied to the entire pipe surface with a gloved hand, brush, or rag. The tape is spirally wrapped with a minimum 2.5 cm (1-inch) overlap. While wrapping, press air pockets out and smooth all lap seams. For additional mechanical protection, an overwrap may be used to increase impact strength and electrical resistance or add ultraviolet protection. Irregular surfaces such as valves, flanges, etc., may require the use of mastic.

3.2.4.2. Hot-applied Wax

Hot-applied wax and wrap provide a coating system for pipeline reconditioning, weld cutbacks, and field patching. Hot-applied wax is a blend of microcrystalline waxes that provide the most desirable corrosion resistance, adhesion, and wetting characteristics. It is hot-applied directly to the pipe surface and is always over-wrapped with a wrapper that depends on the type of application. Its ease of application and forgiving character in difficult applications make it an ideal all-purpose pipe coating. The hot-applied wax and wrap system is environmentally safe and user friendly, compatible with other coatings. It also has a proven field history with over 45 years of use.

After minimal surface preparation, hot-applied wax is flood coated onto the pipe. Hot-applied wax is self-priming. There are no toxic fumes. Outer wrap often is spiral-wrapped over the hot-applied wax to provide backfill protection and to increase dielectric strength. Immediately after application, the pipe can be inspected and backfilled.

3.2.5. Viscous Elastic Coatings

Viscous elastic coatings (VEC) are polar, non-crystalline compounds used with an outer wrap for mechanical protection. VECs are suitable as pipeline rehabilitation coatings due to their adhesion properties on both existing coatings and the protected steel. The coatings are easy to apply and conform around the object easily with minimal surface preparation and tools.

VECs are applied by removing the release liner and placing the adhesive side on the pipe or substrate. It begins with a straight circumference wrap around the pipe. When the initial straight circumference wrap has been completed, the coating is wrapped with slight tension down the pipe, starting on the initial straight wrap. The end wrapping is conducted with a straight circumference wrap. The coating is then smoothed by hand to ensure there are no wrinkles, folds or entrapped air, and the wrap has completely adhered to the substrate. The VECs are to be over-wrapped with a manufacturer-specified PE or composite outer wrap. Outer wrap will be applied under tension at a

50% overlap. Viscous elastic wraps offer immediate adhesion without the need for primer, require minimal surface preparation, and form a homologous, continuous, and self-healing protective layer.

3.2.5.1. Underground Applications

Viscous elastic coatings prevent corrosion on underground pipelines, including wet and irregular surfaces. It requires no waiting or drying time, can be backfilled immediately, and supports CP. VECs are user-friendly, contain no volatile organic compounds, and are non-toxic, non-hazardous, and non-carcinogenic. VECs provide excellent protection to a variety of applications, including couplings, valves, fittings, weld cutbacks, and thermite welds.

3.2.5.2. Aboveground Uses

VECs prevent corrosion on aboveground piping, bridge crossings, vaults, and other straight or irregular surfaces. It is easy to apply, requires no special equipment, is compatible with other coatings, and requires only minimal surface preparation. It is non-toxic and non-carcinogenic, and is composed of inert, non-biodegradable materials. Thus it is essentially unaffected by the elements.

VECs provide long-lasting protection of bridge spans from UV light, weathering, and road salt run-off. VECs completely conform to irregular surfaces, and can be applied with minimal surface preparation that may be found in vaults or meter stations.

3.2.5.3. Pipeline Reconditioning

VECs provide an excellent coating system for pipeline reconditioning, weld cutbacks, and field patching. The VEC system is environmentally safe and user-friendly, compatible with other coatings, and easy to apply. After minimal surface preparation, VECs can be applied to the substrate without primer. VECs have no toxic fumes. Immediately after application, the pipe can be inspected and backfilled.

3.2.6. Concrete Weight Coatings

The wall thickness of pipelines installed onshore may vary with the product transported. A high-pressure natural gas line may require a wall thickness greater than a line transporting a petroleum product. For onshore pipelines, a pipe-wall thickness as thin as 0.25 inch (6.2 mm) may be used. For offshore installation, especially in deep waters, pipelines may require a greater wall thickness, sometimes exceeding 1.0 inch (25 mm) to possess greater strength and to provide negative buoyancy that prevents the pipe from floating (especially pipes with diameters larger than 8 inches (20 cm). Frequently, the heavy-wall pipe is coated with a weight coating such as concrete reinforced with metal wire mesh or concrete with iron-oxide powder for extra weight.

Concrete weight coatings may be applied by a gunite or sprayed-concrete method, compression coat (extruded), or an external wrapped external polyethylene wrap that compresses the concrete.

3.3. Coating Repair and Rehabilitation

Each coating type has to be repaired or replaced for various reasons. The most reliable way to repair or replace coatings is to follow manufacturers' recommended procedures. There are times when original procedures may not be as stringent as those of the user company. More stringent procedures should be followed in these cases. Each environment, each coating type and available repair materials will require more precise repair procedures.

3.3.1. New Coatings

New coatings are typically more easily repaired because the environment at the coating plant is more conducive to proper application and availability of the best repair coatings. Coated pipes are easily damaged during storage, handling, shipping, and construction. The proper repair procedure is more critical after the pipe is in the field.

When the damage or holiday area is located by a holiday detector or visual inspection, the area is marked so it can be easily located later. The proper environmental conditions are monitored and kept as per manufacturers' recommendations throughout the process to ensure the proper conditions exist for the repair coating to perform properly. The area is cleaned with proper solvents, soap, and deionized water rinse (if needed) and abraded with the recommended size grit paper. There are times when the area may be large enough to require grit blasting. The repair coating must be properly cured before the pipe is handled or backfilled.

3.3.1.1. Fusion-bonded Epoxy

FBE coatings are typically repaired using one of the following coating types.

Two-part epoxy. The repairs are usually small enough so the amount of two-part epoxy needed is dispersed through a dispenser gun that has a mixing tube on the end. This mixture is then applied to the prepared area at the wet-film thickness specified and allowed to cure. If there is no dispenser gun, the properly measured amount of each component of the two-part epoxy is thoroughly mixed on a clean surface until there is a consistent color throughout the mixture. This is then applied to the damaged area and allowed to cure.

Tape coating. Tape coatings are used to repair holidays or damaged areas on FBE coating. With tape coatings, a primer (if required) is applied after the surface is prepared. After the primer is cured to the specified amount, the tape is then applied in either a patch that covers the area as specified or in a complete cigarette wrap.

Hot-melt sticks. Hot-melt sticks are misused in many cases and some companies will not allow their use because of failures. The proper technique is critical to their working properly. First, the correct stick for the particular coating type must be used. After surface preparation, the damaged area on the pipe must be heated with a small propane torch to a temperature that will melt the stick when it is touched to the hot surface. Do not melt the stick and allow it to drop onto the damaged area. Allow the pipe to cool to ambient temperature before handling.

3.3.1.2. Multi-layer Coatings

Repair multi-layer coatings using two-part epoxy with a layer of polyolefin. If damage is to the steel in the multi-layer, then the area must be properly prepared and abraded or blasted. A two-part epoxy (same as FBE repair) is applied to the steel. Once cured, it is covered by some type of polyolefin film. This may be in the form of hot-melt stick, a hot-applied film, or a tape. If the damage is only to the outer film, the hot-melt stick or film may be used. Tapes and shrink sleeves are also used for these repairs.

3.3.1.3. Extruded Polyolefin

Cross-head die or side extrusion polyolefin coatings are normally repaired with either tapes or shrink sleeves. Because these coatings expand and contract on the pipe, liquid coatings are not recommended.

3.3.1.4. Coal Tar

Coal-tar coatings are repaired by covering the area with a melted coal tar or coal tar epoxy after surface preparation.

3.3.1.5. Liquid Coatings

Two-part epoxy and other liquid coatings are usually repaired with the same material as the original coating. After surface preparation, the two-part epoxy is mixed and applied according to manufacturers' recommendations and allowed to cure before handling or backfilling.

3.3.1.6. Tape Coatings

Tape coatings, including wax and VEC, are usually repaired using the same material as the original coating. If required, primer is applied after surface preparation. According to the size of the damage, a patch may be applied first, then a cigarette wrap is applied over it to protect the repair during handling and backfilling. No cure time is required.

3.3.1.7. Shrink Sleeves

Shrink sleeves are usually repaired by applying another sleeve over the damaged area. In some cases, tapes may be used.

3.3.2. Coating Rehabilitation

All coatings fail for various reasons. Based on a coating's failure mode, companies may choose to rehabilitate certain sections of the pipe. Through the use of inline inspection data with various coating testing methods in the field, the company can make a decision about what areas need to be re-coated.

If the coating system has proven to be non-shielding to CP when disbondments occur, additional CP can be applied if protection levels are not meeting NACE criteria. Companies can ignore the fact that the coating is shielding to CP and simply add more CP to try to prevent further corrosion. In the latter case, more CP will (most likely) not provide the required protection under these disbonded CP shielding coatings, allowing corrosion to continue. In this case, the only way to prevent further corrosion is to replace the disbonded or failed coating. This is a costly project in many cases, since most of the coating may be disbonded. The decision is critical to future service of the pipeline. Consider the following factors when making the decision:

- safety of the surrounding environment, including population, environment, and company assets
- cost of increasing external corrosion, leaks, and lost profits due to pressure/throughput reduction
- public and regulatory perception
- pipeline life-expectancy and throughput increase if rehabilitation is completed
- continued increased cost of cathodic protection, surveying, and monitoring
- cost of rehabilitation

Once the decision has been made to rehabilitate pipeline coating, a decision must be made about how much of the pipe is to be recoated and what type of coating is best for the rehabilitation. These decisions will be made according to information gathered from the above surveys followed by direct assessments, to confirm actual conditions and the severity of coating damage.

Once the amount of pipe to be recoated is determined, the type of coating used in the rehabilitation has to be decided using the following information:

- For the type of existing coating on the pipeline, how can it be removed and what kind of surface will remain?
- How large is the replacement area? This will help determine the type of coating to be used.
- What type of surface preparation is required for each replacement coating?
- What type of surface preparation is available and at what cost?
- What will the environmental conditions be during the rehabilitation?
- Availability of qualified personnel and equipment for the work.
- Cost of the coating types, even though this is a minor cost when compared to all the other costs of rehabilitation.

3.3.2.1. Liquid Coatings

Two-part epoxy coatings have become one of the most popular coating systems for the rehabilitation of failed coatings. These are usually used for larger jobs, where plural-component spray equipment can be used for the application. For smaller jobs, the two-part epoxy can be applied by roller or brush.

A variety of other liquid coatings have been used for field rehabilitation. Some of these are coal-tar urethanes, polyurethanes, and polyurea.

3.3.2.2. Tape Coatings

The solid film-backed tapes had many issues as discussed earlier, but are still used by some for field rehabilitation. Manufacturers have improved their methods of making the solid film-backed tapes to help reduce some of the problems.

The mesh-backed tapes have become very popular for coating rehabilitation since these do not have the issues with soil stress, disbondment, and CP shielding that the solid film-backed tapes or shrink sleeves have had. These can easily be applied using automated machines or hand wrapsters to provide a more consistent application with proper overlaps and tension.

3.3.2.3. Shrink Sleeves

Shrink sleeves have been used on a limited basis for field rehabilitations. The complicated steps to properly apply shrink sleeves prohibit them from being used for large projects.

3.3.2.4. Other Coatings used for Field Rehabilitation

Coal-tar epoxy, wax-based tape or hot-applied wax, and a few other coating types have been used, but on a more limited basis.

References

Alexander, S. H., Tarver, G.W. (1965) Criteria for Composition and Properties of Hot-Applied Asphalt Pipeline Coatings, American Chemical Society, the 150th National Meeting, Atlantic City, NJ, USA, p. 9:4.

Been, J. (2011) Comparison of the Corrosivity of Dilbit and Conventional Crude, Alberta Innovates Technology Future, Edmonton, AB, Canada.

BrederoShaw 1, AE - Asphalt Enamel Pipeline Coating, Product Data Sheet, Houston, TX, USA (http://www.brederoshaw.com/non_html/pds/BrederoShaw_PDS_AE.pdf).

BrederoShaw 2, Yellow Jacket - High Density Two Layer Polyethylene Coating, Product Data Sheet, Houston, TX, USA (http://www.brederoshaw.com/non_html/pds/BrederoShaw_PDS_YellowJacket.pdf).

BrederoShaw 3, 3LPE - Three Layer Polyethylene Coating, Product Data Sheet, Houston, TX, USA (http://www.brederoshaw.com/non_html/pds/BrederoShaw_PDS_3LPE.pdf).

BrederoShaw 4, FBE - Fusion Bonded Epoxy Powder Coating, Product Data Sheet, Houston, TX, USA (http://www.brederoshaw.com/non_html/pds/BrederoShaw_PDS_FBE.pdf).

BrederoShaw 5, HPPC - High Performance Powder Coating, Product Data Sheet, Houston, TX, USA (http://www.brederoshaw.com/non_html/pds/BrederoShaw_PDS_HPPC.pdf).

Buchanan, R. (2003) Pipeline Coatings & Joint Protection: A Brief History, Conventional Thinking & New Technologies, Rio Pipeline Conference & Exposition, Anonymous Brazilian Petroleum and Gas Institute - IBP.

Canadian Standards Association (2010) Plant-Applied External Polyethylene Coating for Steel Pipe, CSA standard Z245.21-10, Toronto, ON, Canada.

Dickerson, J.G. (2001) FBE Evolves to Meet Industry Need for Pipeline Protection, *Pipe Line & Gas Industry*, 84, 67-74.

Goldie, B. (2010) Developments in Pipeline Protection Reviewed, in: *Protecting and Maintaining Transmission Pipelines - A JPCL eBook*, Technology Publishing Company, 9-12.

Guan, S.W. (2011) Challenges and New Developments in Pipeline Coating Technology, Oil & Gas and Petrochemical and Gas Processing Industries in Asia, (Petromin Pipeliner).

Howell, G.R., Cheng, Y.F. (2007) Characterization of high performance composite coating for the northern pipeline application, *Prog. Organ. Coat.* 60, 148-152.

Johnson, W., Chitkara, N.R. (1973) Corrugated plate formed by side extrusion with two coaxial rams moving at different speeds, *Inter. J. Mech. Sci.* 15, 199-210.

Kehr, J.A., Enos, D.G. (2000) FBE, a Foundation for Pipeline Corrosion Coatings, Corrosion'2000, paper no. 00757, NACE International, Houston, TX, USA.

Kehr, J.A., Hislop, R., Anzalone, P., Kataev, A. (2012) Liquid Coatings for Girthwelds and Joints: Proven Corrosion Protection for Pipelines, in: *Protecting and Maintaining Transmission Pipelines - A JPCL eBook*, Technology Publishing Company, 23-32.

Keller, A. (1956) Studies of orientation phenomena in crystallizing polymers, *J. Polymer Sci.* XXI, 363-379.

McManus, J. J., Pennie, W.L., Davies, A. (1966) Hot Applied Coal Tar Coatings, *Industrial & Eng. Chem.* 58, 43-46.

Munger, C.G. (1999) *Corrosion Prevention by Protective Coatings*, NACE Publishing, Houston, TX, USA, pp. 520.

NACE (2011) *Coatings in Conjunction with Cathodic Protection*, NACE International, Houston, TX, USA.

NACE/SSPC (2010) *Near-White Metal Blast Cleaning*, NACE No. 2/SSPC-SP 10, NACE International/ The Society for Protective Coatings joint standard, Houston, TX, USA.

Neal, D. (1998) *Fusion-Bonded Epoxy Coatings: Application and Performance*, NACE, Houston, TX, p. 63–65.

Norsworthy, R., Hughes, C., "Proven Protection", World Pipelines (October 2007)

Norsworthy, R., *Fail Safe Tape System used in conjunction with Cathodic Protection*, Materials Performance, June 2004, pp. 34 - 38.

Pratt, J. K., Mallozzi, M., D'Souza, A (2011) Advances in Damage Resistant Coating Technology, Corrosion'2011, paper no. 11031, NACE International, Houston, TX, USA.

Romano, M., Dabiri, M., Kehr, A. (2012) The ins and outs of pipeline coatings: Coatings used to protect oil and gas pipelines, in: *Protecting and Maintaining Transmission Pipeline - A JPCL Book*, Technology Publishing Company, Pittsburgh, PA, USA.

Schad, M. (2013) Corrosion protective coatings: rehabilitation of buried steel pipelines, *Pipelines Inter.* (9) 54-55.

Shaw Pipe (2010) *Pipeline Coating Solutions Technical Literature 2010*, Toronto, ON, Canada.

Shaw Pipe (2012) *Pipeline Coating Solutions, Technical Literature 2012*, Shaw Pipe TL R161008, Toronto, ON, Canada.

Singh, P.J., Cox, J.J. (2000) Development of a cost-effective powder coated multi-component coating for pipelines, Corrosion 2000, paper no. 762, NACE International, Houston, TX, USA.

Singh, P.J, Haberer, S., Gritis, N. (2005) New developments in high performance coatings, 16th International Conference on Pipeline Protection, Cyprus.

Singh, P.J, Williamson, A.I. (1999) Development of a high performance composite coating for pipelines, NACE Northern Area Conference, Calgary, AB, Canada, March 1999.

Tracton, A.A. (2010) *Coating Technology Handbook*, CRC Press, Ch. 54, 54-1.

Trossarelli, L., Brunella, V. (2003) *Polyethylene: Discovery and Growth*, Dipartimento di Chimica IFM, Via Giuria 7, 10125 Torino, Italy.

Vincent, L.D. (2001) Ambient Condition Effects on Pot Life, Shelf Life, and Recoat Times, *Mater. Perf.* 40, 42–43.

Willbourn, A.H. (1983) *The Origin and Discovery of Polythene*, Lecture at the Golden Jubilee Conference, POLYETHYLENES 1933-1983, London, UK, 8-10 June.

4 Coating Failure Mode and Effect Analysis

4.1. Coating Failure Modes and Mechanisms

Pipeline coatings can fail at various stages: In manufacturing, application, transportation, or pipeline construction and servicing. While this book focuses on coating failures in field use, the mechanisms causing coating failure at other stages are briefly described below.

Design. The pipeline design, whether it is used above ground or underground, can greatly influence the coating system's performance and longevity. For above-ground pipelines, the coating may degrade or fail through UV radiation when the coated pipe is exposed to direct sunlight. Abrasion can also occur on above-ground coating, which is vulnerable to direct contact by workers and machines. Moreover, structural supports may damage the coating locally, especially where there are ground settlements. For underground pipelines, the coating's primary function is to block the pipe steel for corrosion protection from the soil environment. The soil chemistry, moisture, resistivity, microbial activity, and CO_2 gas can impact the coating integrity and affect its performance and failure modes. The effect is further enhanced by cyclic dry-wet alterations and temperature changes. The CP system may facilitate coating disbondment, especially at pinholes and holidays. Impact caused by excavation equipment can damage the coating and even the pipe.

Manufacturing. Manufacturing coated pipes in plants includes two steps (i.e., the manufacturing of the pipe [both seamless and welded pipes], and the application of the coating). The coating can be affected by the pipe manufacturing step, since various manufacturing defects (such as rolling defects and weld defects) can affect the coating application. Moreover, a number of factors introduced during coating application can cause coating degradation and failure. For example, inadequate surface preparation and cleaning may leave contaminants like slugs that reduce coating adhesion; abrasive cleaning may be inadequate for providing the required surface profile patterns for coating to adhere tightly to the substrate; and the environment where the coating is applied may not be well-controlled (as in high humidity, which affects the coating application).

Transportation, storage, and handling. Coatings can be damaged during transportation and handling. Pipe transportation, loading, and unloading can cause dents, abrasions, gouges, and other defects. Pipe handling equipment (such as forklifts) may damage the pipe and/or the coating. Moreover, the pipe storage environment may have dirt, dust, chemicals, and moisture, or even exposure to high-UV light. All of these environmental conditions pose threats to coating's integrity.

Construction. Backfill materials may contain rocks, which can damage the coating. Field welding may generate undetected weld defects and improper surface conditions due to labor skill, resulting in poor coating adhesion to the steel substrate.

While pipeline coatings can fail in the stages mentioned above, varied mechanisms result in coating failure in the field. These failures can be due to poor coating materials and formulation, inappropriate substrate surface treatment, abnormal operating conditions, and corrosive service environments. As a result, a wide variety of coating failure modes are encountered during pipeline service.

4.1.1. Coating Disbondment

Among various coating failure modes, adhesion-related failures are most common in the field. Adhesion-related failures can cause catastrophic consequences and sudden pipeline failure. Disbondment occurs when a coating loses adhesion to the substrate metal due to a number of reasons described below.

Poor surface preparation. If the metal substrate is not prepared properly, the likelihood of disbondment increases significantly, regardless of how well the coating has been applied, or how thoroughly subsequent coating processes have been controlled. Indeed, poor surface preparations have always been the major cause of coating disbondment. Surface contaminants (such as dirt, greasy, moist, or rusting) can reduce the coating's bonding to the metal. Thus, factors affecting the substrate's surface cleanliness can change (usually degrade) coating adhesion, causing disbonding. Moreover, the surface profile pattern and roughness, which can be described as the number of peaks and valleys per unit length, affect the adhesion. Generally, the required standard for surface preparation depends on the coating to be applied. The surface must be clean and free from oil or grease. It should be solvent-washed or cleaned with a suitable emulsifier. In most cases, the surface is to be dry and at least 37 °F (3°C) above the dew point. Furthermore, coatings that chemically reacted with the pipe steel require an abrasive blast-cleaned surface of the substrate [Jackson, 2015].

Inadequate processing procedure. A poorly applied coating is much more likely to disbond from the metal substrate than is a well-applied coating. Factors that reduce coating adhesion include various scenarios (e.g., the substrate is too hot or too cold, the coating is not at the right temperature, the coating is not mixed at the right ratio, the coating is not applied with intimate contact to the substrate, or coating was given inadequate curing or cooling time). Sometimes, solvents remain part of a coating for a long time and become entrapped in the coating after a film forms on the substrate. When temperature increases, these solvents produce vapors. When the vapor formation rate becomes greater than the vapor diffusion rate, blistering may occur at the substrate/coating interface or in the coating's interior. Blisters may rupture when internal pressure is sufficiently high. As a result, coating disbondment occurs. If blisters break inside the coating, the coating experiences delamination. In addition to the failure modes mentioned above, coatings can also fail in the form of chalking, cracking, discoloration, and similar phenomena.

Temperature variations between coating and metal surface. Metal surfaces are generally colder compared to coatings. Hence, if a permeable coating is used in a warm environment containing moisture or water (e.g., warm aqueous solution), the moisture or water passes through the coating and reaches the steel surface. As the steel surface is cooler than the vapor, it condenses beneath the coating and results in local disbonding.

Improper selection of primer. Problems associated with inappropriate material selection are more likely to cause disbondment. Multi-layered coatings usually use FBE as primer for its strong adhesion. However, thermo-oxidative degradation can occur on FBE and reduce its interfacial adhesion. Prior to 2000, most FBE coatings had the T_g around 212°F (100°C), and the maximum service temperature for external FBE coating is specified at 140°F (60°C). The FBE can become brittle under long-term thermal aging. The effect of thermo-oxidative aging, heating, and oxidization on FBE's adhesion to steels indicates that thermally embrittled FBE is prone to disbonding from the steel.

Soil stresses. In geographically unstable regions (such as on slopes or in earthquake-prone areas), pipelines may incur significant longitudinal stress or strain due to ground movement. In these circumstances, the applied coating typically forms folds and ripples that eventually tear and disbond from the substrate steel.

4.1.2. Blistering

Blistering is another type of coating failure attributed to weak adhesion of the coating to the substrate. Generally, blistering is common when a coating is immersed in water or aqueous solutions. It can also occur when the service environment is highly humid or when there is frequent water condensation on the coating surface.

When a coating absorbs gases or liquids that accumulate inside the coating or at the coating/substrate interface, pressures are generated locally. When the pressure is sufficiently high and exceeds the adhesion of the coating to the substrate or the internal cohesive force of the coating, the coating can be stretched and form hemispherical, blister-like shapes. This process is called blistering. When the coating's mechanical strength is greater than the local pressure, blisters become bigger. When internal pressure exceeds the coating strength, the blister breaks and damages the coating. Substrate materials, such as pipeline steels, become directly exposed to the environment with damaged coatings.

Various factors contribute to the formation of liquid or gas blisters beneath or within the coating. One of the most important factors is a coating's moisture vapor transfer rate, i.e., the rate at which water molecules pass through the coating's intermolecular spaces. The moisture vapor transfer rate is a characteristic of coating formulation. Moisture tends to accumulate at the coating/substrate interface in areas with poor adhesion. Vapor pressure is created in these areas due to moisture-vapor accumulation, causing blister formation. Generally, coatings with a smaller moisture-vapor transfer rate are more blistering-resistant. The coating adhesion is critical to blistering-induced failure. A coating with excellent adhesion usually shows little tendency to for blistering, even with a high moisture-vapor transfer rate.

Osmotic blistering. Osmosis occurs when moisture vapor passes through a coating film from a less-concentrated water solution side to the concentrated-solution side. Blistering caused by water movement due to osmosis is classified as osmotic blistering, one of the most commonly occurring types of blisters on coatings in the field. Generally, osmotic blistering requires three prerequisite conditions to occur (i.e., a water-permeable coating film, a comparatively impermeable substrate, and a concentration gradient).

When the coated metal structure is used in aqueous environments and the coating is permeable to water, moisture penetrates through the coating from the exterior towards the substrate on the inner side of the coating. The moisture vapor passing across the coating can generate an osmotic pressure. If the coating adhesion to the metal substrate is insufficiently strong and if the vapor pressure reaches and exceeds a certain level, blisters form under the coating. If the coating's adhesion strength is greater than the vapor pressure, blistering will not occur. Thus, the sufficient coating adhesion strength may avoid blister formation, including osmotic blistering.

Osmotic blistering can occur in various inorganic solutions (such as chlorides and sulfates), and also in organics (such as sugar). These aggressive media can corrode metal under the coating and the coating blistering. It was found [Munger, 1994] that ferrous-sulfate solutions can produce a large quantity of fine blisters, whereas sodium chloride results in fewer but larger blisters.

Sometimes, osmotic blistering may occur at small coating defects (e.g., pinholes, pores, etc.), where coating blistering and corrosion of substrate metals at defects are interrelated.

Electro-endosmosis blistering. Electro-endosmosis is the process of water molecules penetrating a coating film, driven by an electrical current in the direction of the electrical pole with the same charge as that of the coating. Most coatings used in industry are negatively charged, which can cause water molecules to move toward the negative pole (i.e., pipe steel) when a steel pipeline is under CP potential and becomes a source of negative electrons. If the coating has a poor adhesion locally, moisture movement due to electro-endosmosis produces vapor pressure in areas of poor adhesion, which eventually results in blister formation. It has been reported that one of the key reasons that pipeline coatings suffer from blistering is the CP application on the pipeline. Generally, cathodic potentials exceeding -1.0 V (CSE) cause blistering problems at locations of poor coating adhesion.

Cathodic blistering. During corrosion of metals, the reduction of dissolved oxygen dominates the cathodic reaction. An alkaline environment can be generated locally with the production of hydroxyl ions that elevate solution pH:

$$O_2 + 2H_2O + 4e^- \rightarrow 4OH^- \qquad (4\text{-}1)$$

Organic coatings used for pipelines are vulnerable to alkali attack, resulting in disbonding of the coating and the generation of blisters. Usually, the anodic reaction (i.e., the oxidation of iron) occurs at the center of the disbondment, and the cathodic reaction (4-1) occurs at the steel/coating boundaries. Cathodic blistering requires that CP current permeates the coating.

Thermal gradient-induced blistering. Thermal gradient-induced blistering, usually referred as cold-wall effect, results from water condensation on the substrate that is produced by a thermal gradient across coatings. This blistering is commonly observed on interior surfaces of cold, coated tanks containing warm liquid.

4.1.3. Pinholes and Holidays

Various defects or damages may exist in pipeline coatings. Microscopic defects in the form of a pinhole or holiday could be introduced into the coating during the manufacturing process. Moreover, mechanical coating damage is almost unavoidable during coated pipe transportation and pipeline construction. It has been reported [Masilela and Pereira, 1998] that, on an investigated 31-mile (50-km) gas pipeline coated with FBE, the portion of mechanical damage can be up to 80% of the whole length.

A holiday is a relatively big break in the coating system, where bare steel is directly exposed to the soil environment. Pinholes are small coating defects. However, there is no a clear size boundary between a pinhole and a holiday. The geometric difference between a pinhole and a holiday can result in distinguishable corrosion behavior occurring on the steel at the base of the two types of coating defect. While pinholes are primarily generated during coating manufacturing, holidays can be introduced to coatings by manufacturing (less common), application, and pipeline construction (more common). The pinhole or holiday itself becomes a corrosion cell that results in local corrosion as well as initiation of coating disbonding. Therefore, careful inspection after coating application is very important for a long and defect-free service life of the coating.

4.1.4. Cracked and Missing Coatings

The coating may experience frequent damage during pipeline installation. Typically, the coating cracks during in-field bending of pipes, penetration of pipe supports into the coating due to unsuitable design and storage conditions, and use of unsuitable slings in pipe handling. Coating cracks and/or even locally missing coatings can also occur while pipeline is buried in a trench. Causes include backfill impact, handling abrasion due to pipe jacking, penetration of rocks or stones due to the weight of the pipe and/or backfill, shearing action of soil due to compaction and subsidence, and ground movement due to geotechnical factors such as unstable slopes and earthquake. Moreover, a coating can be cracked during pressure-testing of pipelines.

Coatings can also be cracked by internal stresses, both tensile and compressive stresses caused by wet-dry cycling of the coating. As a coating absorbs water, it swells and generates compressive residual stress. When the coating dries, it contracts and generates tensile residual stress. The compressive and tensile stresses reduce the coating's cohesive strength and its adhesion to the substrate. Coating stress is a dynamic phenomenon, depending on water uptake and desorption [Sato, 1980]. During the dynamic stressing process, some degrees of permanent creep can occur on the coating, which is attributed to breaking and reforming valance associations in the epoxy polymer. Remarkably, hygroscopic stresses affect coating performance. If a coating forms high levels of internal stress during curing, other stresses generated during water absorption or desorption can lead to coating cracks or delamination.

When a coating cracks and is extensively damaged, the mode called "coating missing" exposes uncovered pipelines to soil. This can happen due to soil sliding, floods, any natural disaster, or during soil excavation. Moreover, inappropriate coating strength makes the coating incapable of withstanding external forces (e.g., backfill impact), resulting in missing coating.

4.1.5. Material Degradation in Service Environments

Ultraviolet breakdown. It is a commonly observed phenomenon that coated structures exposed to direct sunlight suffer from aesthetic changes (such as yellowing, color loss, chalking, gloss reduction, etc.). Moreover, a coating may experience chemical breakdown and degradation of mechanical properties, causing increased embrittlement, hardness, and internal stress, loss of film thickness, disbonding from the substrate or delamination among coating layers, changed solubility and cross-link density, and decreased permeation in barrier properties. All damages as listed are due to UV components with a wavelength from 400 nm to 100 nm, which possess sufficient energy to break chemical bonds. While the short wavelength part of the UV has the most destructive power, it cannot penetrate deep enough and only causes damage to the surface layer of coatings. Longer wavelengths can penetrate deeper inside the coating, but cause less damage. This leads to coating inhomogeneity, where top layers are more highly cross-linked than the bulk of the coating.

The interaction of coatings with the UV includes UV reflectance, UV transmittance, and UV absorption. In general, reflectance and transmittance do not affect the coating's service life, but UV absorption can lead to chemical destruction of coating molecules. Sunlight can be absorbed by pigment, binder, or additives contained in the coating. Damages come mainly from UV absorption by non-pigment components of the coating (i.e., the polymeric binders or additives). Absorbed UV energy can damage the polymeric material and break existing bonds.

Chemical breakdown. Coatings can be contaminated by water and electrolytes that contain Cl^- or SO_4^{2-} ions. At the same time, water can be a solvent for some coating additives, resulting in additive dissolution or leaching (out of the coating film). It was proposed [Forsgren, 2006] that each coating has a tolerance to critical relative humidity (RH). Above the critical RH, water condenses on OH^- groups of the polymer, breaking inter-chain hydrogen bonds and displacing adsorbed OH^- groups from the substrate. This can result in the loss of coating adhesion to the substrate.

Chemical degradation. Chemical degradation refers to coating breakdown induced by exposure to chemical contaminants in service environments. Coatings may deteriorate from chemical degradation in high salinity areas if they do not have adequate resistance to the conditions. The degradation process accelerates with heat. An elevated temperature increases the amount of chemical ions exchanged between the coating and the environment, reducing coating resistance. Although polymers used in modern coatings have a good resistance to acids and salts, the coating also contains a number of additives that are usually vulnerable to chemical attack. For example, many coatings contain light stabilizers to improve UV resistance. However, the performance of these stabilizers is reduced by acids and pesticides. When this occurs, the chemical exposure makes the coating vulnerable to UV breakdown. There have been proven links between industrial pollution and coating damage [Forsgren, 2006].

Thermal degradation. Coatings may degrade due to incorrect temperatures in actual operating environments. Pipelines operating at too-high temperatures may cause the coating to gradually deteriorate and break up. In extremely cold temperatures, the coating may have insufficient flexibility and crack when the pipe is handled or when it moves on the ground due to thermal contraction. Moreover, ambient temperature changes can alter stresses in the coating/substrate system, mechanical properties of the coating, and diffusion of electrolyte and water through the coating. At slightly elevated temperatures, the coating can harden and eventually crack. Some polymers crack at higher temperatures, but they weaken when temperature increases even slightly. Differences in coefficients of thermal expansion between coating and the substrate metal also cause thermal stress. Furthermore, at elevated temperatures where the T_g of polymers used in the binder is exceeded, the polymer exists in a rubbery state; below T_g it is in a glassy state. Thus, when coatings are used near the T_g range, the binder can experience changes in the transition from glassy to rubbery state. It was found [Forsgren, 2006] that, above T_g, polymer chain segments undergo Brownian motion, leading to a high diffusion rate inside the polymer. Pores as small as 1 to 5 nm are generated within the binder matrix. The permeation rate increases through these small pores as temperature increases, causing reduced coating resistance in the service environments.

4.2. Coating Failures and Cathodic Protection Performance

Buried pipelines are protected from corrosion attack by coating and CP. The principle of this strategy is very straightforward (i.e., the coating provides the first line of barrier to separate the pipeline from soil environments). Once the coating degrades or fails, the CP continues to provide essential pipeline protection. Obviously, the coating and CP performances are highly related. Coatings should be considered when the required CP current is too high to protect a structure. A high-performance coating can reduce the CP current requirement for corrosion protection. The amount of required CP current depends on coating properties and the amount of coating damage. When the coating fails (e.g., the coating is disbonded from the substrate), the CP current should penetrate through the coating to reach the substrate for corrosion protection.

4.2.1. Principle of Cathodic Protection

Cathodic protection is a technique to control corrosion of a metal structure by making it the cathode of an electrochemical cell [Peabody, 2001]. The electrochemical cell for a corrosion reaction includes an anode (i.e., the segment of the metal structure where oxidation (or corrosion) of the metal occurs), a cathode (i.e., the segment of the structure where the electrochemical reductive reaction occurs), and an electrically conductive pathway such as an electrolyte. The CP is an electrochemical technique enabling change of the metal, which corrodes in service environments in the absence of

CP or other protective measures, from anode to cathode, thus preventing occurrence of corrosion. Actually, the CP is also a process to cathodically polarize the target structure.

The CP is probably one of the most important measures used for corrosion control for engineering structures. Principally, a metal surface that is under CP can be maintained in a corrosive environment without deterioration within an indefinite time if the CP is appropriately maintained [Revie and Uhlig, 2008]. There are two types of CP systems (i.e., impressed current cathodic protection [ICCP] and sacrificial anode cathodic protection [SACP]).

Impressed current cathodic protection. The ICCP system uses an external source of electrical power, which is often a transformer-rectifier connected to AC power, where its negative terminal is connected to the structure to be protected. Its positive terminal is connected to an auxiliary electrode (so-called anode), which is usually made of iron, graphite, or metal oxides, located some distance away from the protected structure. The current flows from the auxiliary electrode through the electrolyte to the protected structure. In ICCP, the applied voltage is not critical, and it just needs to supply an adequate current density to all parts of the protected structure. The current source is usually a rectifier supplying low-voltage direct current (DC). The main advantage of the ICCP system is its automatic control feature, which continuously monitors and varies the current required for corrosion protection of the structure. However, ICCP systems require high inspection and maintenance costs, and suffer from a high risk of causing both stray current interference and over-protection.

Sacrificial anode cathodic protection. A simple method for applying CP is to connect the metal structure to be protected with another metal that corrodes more easily in the environment, and thus acts as the anode. When the two metals are electrically connected to each other in the environment, electrons flow from the more active (more easily corroded) metal to the other (metal to be protected) due to the difference in their electric potentials. When the more active metal (anode) supplies current, it dissolves (corrodes) into ions, which go to the electrolyte. At the same time, the anode produces electrons, which are received by the less-active metal (i.e., the protected metal structure) through the metallic connection. As a result, the protected metal is negatively polarized, and hence, protected against corrosion. For the SACP to work, the anode must possess a more negative electrode potential than that of the cathode (i.e., the structure to be protected). Galvanic series of metals can be used to select the anode metal for SACP application [Roberge, 1999]. The advantages of the SACP system include that it does not need an external power source, it is easy to install, overprotection is unlikely, and inspection and monitoring is simple for trained personnel. SACP has a few disadvantages, including a limited current capacity based on the mass of the anode, and the ineffectiveness of corrosion protection in high-resistivity environments.

4.2.2. Conjunction of Coating and CP on Pipelines

As stated, buried pipelines are protected from external corrosion in soil environments by both coating and CP. If the coating possesses excellent properties (such as water resistance, dielectric strength, adhesion, etc.) and does not degrade in the service environment, it can provide sufficient pipeline protection and at the same time, reduce the CP current requirement remarkably. The CP is functional for corrosion protection at coating defects (such as holidays). Coatings with poor resistance properties will require a greater CP current output for corrosion control.

To estimate the total current required for CP design, the percentage of the coated surface area (or the percent of the damaged area where bare steel is directly exposed to soil electrolyte) must be known for a metal structure. If the coating is intact and the pipeline has a 100% flawless coating, there would be no need for CP as the steel has zero contact with the soil environment [Bollinger, 2015]. In reality, coating damage is inevitable. Damage includes pinholes and holidays, scratches, damages by backfills, etc. Thus, an estimate of the possible coating damage percentage is required to calculate the amount of CP current needed for adequate pipeline protection.

Coating damage factors are often used when the amount of equivalent coating damage can be estimated. When the coating damage factor is zero, the coating is completely electric insulating, decreasing the CP current requirement to zero. If the coating damage factor is 1, the coating loses its ability for corrosion protection, and there is no current-reducing effect [NACE, 2002].

The coating damage factor, f_c, is a function of the coating properties, operational parameters, and time. Generally, the f_c can be estimated by [DNV, 2010]:

$$f_c = a + b \times t \tag{4-2}$$

where t is service time of the coating (year), and a and b are constants that depend on coating properties and the environment. The initial coating damage factor, $f_{c,i} = a$, is used to calculate the initial CP current demand for coated structures.

The mean and final coating damage factors, f_{cm} and f_{cf}, respectively, can be calculated using a CP design life, t_f, [DNV, 2010]:

$$f_{cm} = a + b \times (t_f/2) \tag{4-3}$$

$$f_{cf} = a + b \times t_f \tag{4-4}$$

The CP system can be designed based on the coating damage factor to adjust current demands during progressive coating deterioration. The coating can disbond from the substrate steel in-service, resulting in the possibility that the coating shields the CP current from reaching the steel surface.

Besides coating damage factors, all coatings have a dielectric strength, which varies with the coating type. The dielectric strength of a coating is its ability to act as an electrical insulator between the pipeline and the soil. Even with a 100%-intact coating, certain amounts of CP current can pass through the coating with a low electric strength, thus affecting the CP system's design life.

Regardless of the coating type, it is important to check the entire pipeline for coating imperfections. This ensures a minimum number of coating holidays and defects, and a much better performance from the designed CP system.

4.2.3. CP Shielding by Coating Failures—Part I. The Problem

When a coating disbonds from the substrate at small faults, such as pinholes or holidays, the CP current can become shielded, either fully or partially, to reach the disbonding crevice, especially at the crevice bottom. As a result, the CP is unable to protect the area that is in a corrosive environment. This is called "CP shielding." Another scenario encountered in reality that causes CP shielding is the disbondment of a defect-free coating due to either an inadequate coating application process or the coating's lost adhesion to the steel substrate during service. For example, spirally wrapped tape coatings were frequently found to disbond over pipeline welds. In this situation, the CP shielding is attributed to the coating property [Pikas, 1996]. The coating itself blocks the CP current from reaching the substrate metal. Generally, non-shielding coatings are those that do not prevent distribution of CP current to the steel substrate through the disbonded coating [Jack et al., 2002].

It has been reported [Norsworthy, 2008] that up to 85% of all pipeline external corrosion is under disbonded CP-shielding coatings. Thus, the coating used for pipelines should be compatible with CP, as advised by a number of national and international codes and standards. In other words, pipeline coatings should not be cathodic shielding. The relevant regulations from the Code of Federal Regulations (CFR) of the U.S. Department of Transportation (DOT) are 192 [U.S. Code of Federal Regulations Title 49, 2012a] and 195 [U.S. Code of Federal Regulations Title 49, 2012b], which apply to gas and liquid pipelines, respectively. In particular, one section in CFR 192 states that, under certain conditions, the pipe must be protected against external corrosion by a non-shielding coating. It further indicates that the type of non-shielding coating that should be considered is FBE and, in some cases, liquid epoxy.

However, all coatings, including pipeline coatings, are somewhat electrically resistant.

It has been acknowledged [Munger, 1994] that water resistance is one of the most important characteristics possessed by a good coating. A highly water-resistant coating simply means a combination of properties, including a high resistance to water absorption, a low moisture vapor transfer rate, a high resistance to osmosis and electroendosmosis, a high dielectric strength, and resistance to ionic passage. Apparently, a good coating, based on its property requirement, will be CP shielding.

Generally, whether a coating is ranked as "good" depends on whether the coating possesses a combination of good properties. In addition to water resistance, good coatings should be excellent in other properties, such as adhesion, chemical resistance, flexibility, weather resistance, and resistance to dirt pickup and bacterium growth. In particular, proper adhesion is usually the key for a coating to retain its integrity during the service period. For pipeline coatings, the CP compatibility must be considered with the coating's adhesion to the steel substrate. If a coating's adhesive property is ignored, it does not make sense to discuss its CP shielding or non-shielding behavior. Actually, corrosion and SCC do not occur on pipelines when the coating maintains a proper adhesion to the substrate, regardless of its CP permeability. Therefore, whether a coating is compatible with CP matters only when the coating is disbonded [Cheng, 2013].

Finally, the anti-corrosion and high electrical insulation of a coating are two different properties requiring diverse material selections [Muncaster and Perrad, 2015]. Generally, an electric insulator based on current technology cannot be expected to provide the same anti-corrosion performance as a high-performance coating and vice versa. Thus, the functions of high-strength electric insulation and anti-corrosion control are not presently deliverable in the same coating product.

4.2.4. CP Shielding by Coating Failures—Part II. Defect-free Coatings

The CP permeability through a defect-free coating can be measured with the experimental setup shown in Figure 4-1, where two test chambers are separated by a defect-free coating film. A carbon rod is placed in the top chamber and used as the auxiliary anode. The steel electrode is installed in the bottom chamber. CP potentials are applied through an external DC power supply. A gap between the steel electrode and the coating is created to simulate a disbondment crevice, and the disbonding thickness can be adjusted by using an instant adhesive. The detailed procedure for applying the adhesive tape to create the disbonding crevice with varied thicknesses was previously described [Kuang and Cheng, 2015a].

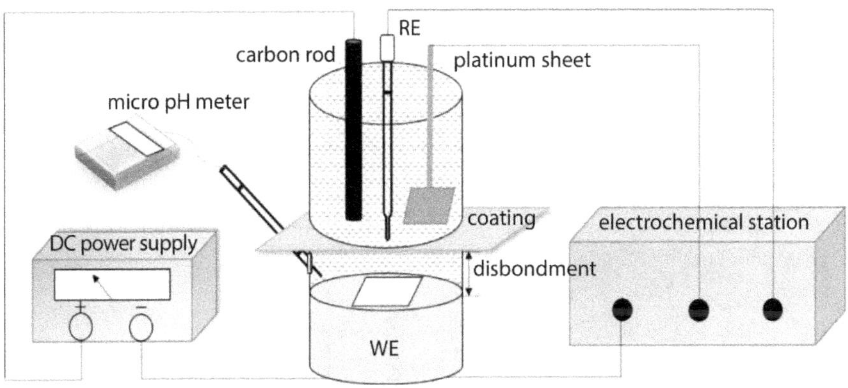

Figure 4-1. Schematic diagram of the experimental setup to measure the permeability of a coating to CP, where WE and RE refer to working electrode and reference electrode, respectively [Kuang and Cheng, 2015a].

Figure 4-2 shows the potential of X65 steel in a 0.01 M $NaHCO_3$ solution trapped under the simulated disbonded HDPE and FBE coatings at CP potentials of -0.85 V(CSE) and -1.00 V(CSE), respectively. Figure 4-2a shows that, even under the CP potential of -0.85 V(CSE), the potential of the steel under HDPE is positive in the beginning and shifts negatively with time. The steel is at about -0.065 V(CSE) only after 30 days of immersion in the solution. Obviously, the CP current could not permeate the HDPE to cathodically polarize the steel. For FBE coating, the potential of the steel is much more negative at individual times, as shown in Figure 4-2b. For example, the potential of the steel reaches -0.35 V(CSE) after 30 days, compared to −0.065 V(CSE) for HDPE. Thus, the FBE is more compatible with CP in terms of the CP current permeating the coating. Although the CP current can permeate the FBE, it does not fully penetrate the coating within the testing time period. When the applied CP level becomes more negative, up to -1.00 V(CSE), the potential of the steel shifts rapidly in a negative direction, as shown in Figure 4-2c. The potential reaches to -0.70 V(CSE) after 10 days and to -0.75 V(CSE) after 30 days of CP application. Thus, a more negative CP level is helpful for CP-current permeation through the coating.

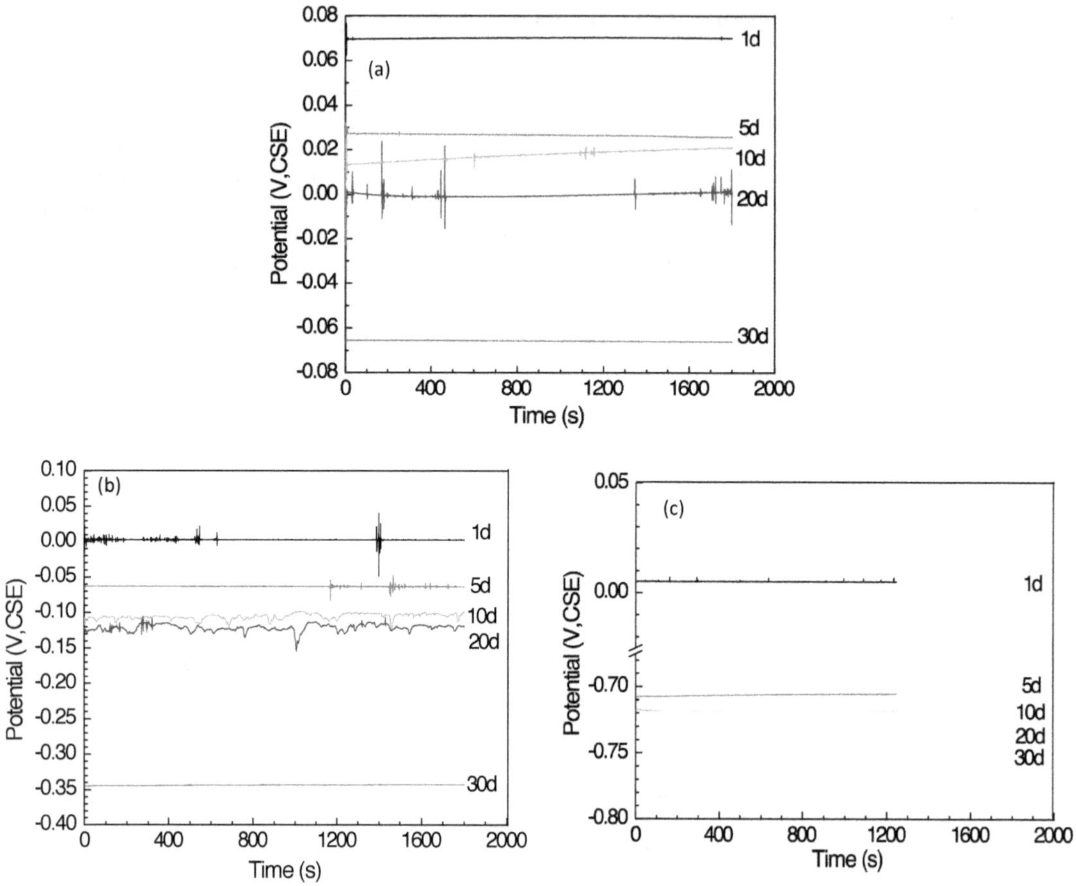

Figure 4-2. The potential of X65 steel in 0.01 M $NaHCO_3$ solution trapped under the simulated disbonded HDPE and FBE coatings under CP potentials of −0.85 V(CSE) and −1.00 V(CSE), respectively: (a) HDPE, −0.85 V(CSE); (b) FBE, −0.85 V(CSE) and (c) FBE, −1.00 V(CSE) [Kuang and Cheng, 2015a].

Figure 4-3 shows the potentiostatic current measured on the steel in 0.01 M $NaHCO_3$ solution trapped under the disbonded HDPE and FBE coatings under CP potentials of -0.85 V(CSE) and -1.00 V(CSE), respectively. Figure 4-3a shows that anodic current densities are observed on HDPE-separated steel when the CP potential of -0.85 V(CSE) is applied. The HDPE coating behaves like an ideal capacitor. The measured anodic current density is associated with the capacitive behavior of the coating. It indicates that the applied CP cannot penetrate the HDPE. For FBE coating (Figure 4-3b), the anodic current density obtained on the first day is from the coating's capacitive behavior, and the CP has yet to penetrate the coating. With increasing time, the cathodic current density is recorded, indicating that CP current penetrates the coating to cathodically polarize the steel. When a more negative CP potential of −1.00 (CSE) is applied, the CP permeation is enhanced, shown by the larger cathodic current density compared to that measured at the CP of -0.85 V(CSE) at individual times, as shown in Figure 4-3c. Moreover, it is found that, even at the CP potential of -1.00 V(CSE), the anodic current density is obtained after one day of testing. Thus, although the FBE is CP-permeable, it will take time for CP current to penetrate.

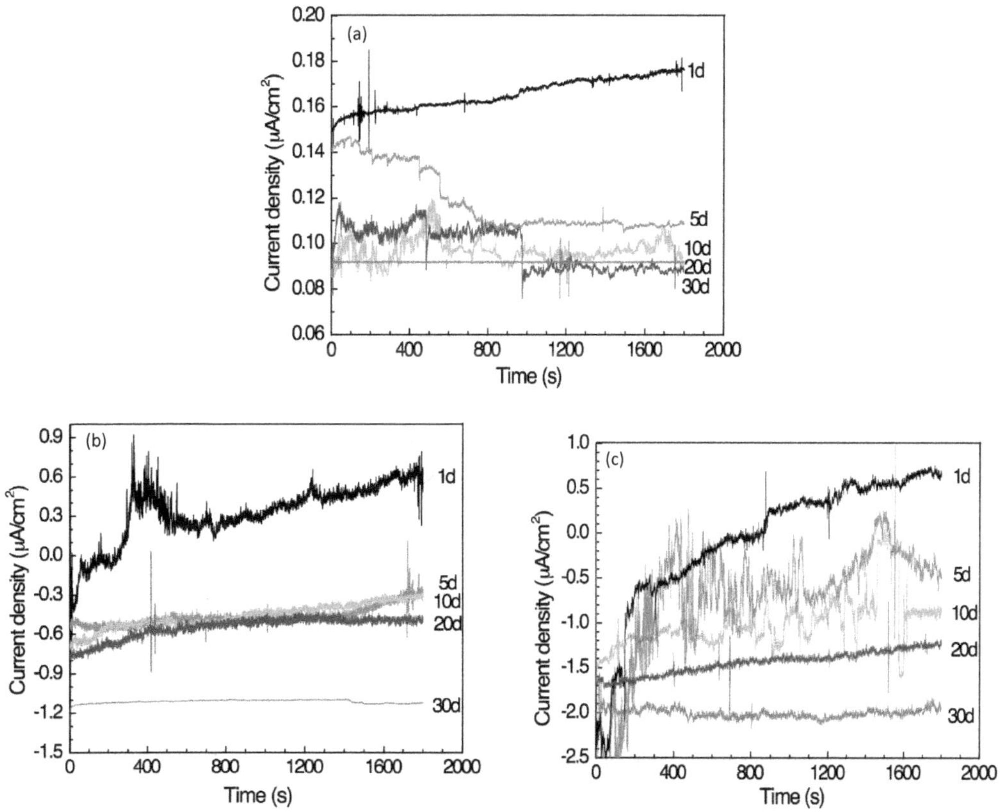

Figure 4-3. Potentiostatic current measured on steel in 0.01 M $NaHCO_3$ solution trapped under the disbonded HDPE and FBE coatings under CP potentials of −0.85 V(CSE) and −1.00 V(CSE), respectively: (a) HDPE, −0.85 V(CSE); (b) FBE, −0.85 V(CSE) and (c) FBE, −1.00 V(CSE) [Kuang and Cheng, 2015a].

Figure 4-4 shows the time dependence of the solution pH under HDPE and FBE coatings at −0.85 V(CSE) and −1.00 V(CSE) of CP potentials, respectively. The pH of the solution under HDPE fluctuates between 7.5 and 8.0 within the 30 days of the testing period, as shown in Figure 4-4a. For FBE coating, the pH of the solution increases continuously with time, and is up to 9.3 after 30 days (Figure 4-4b). When a more negative CP potential of −1.00 V(CSE) is applied to the steel under the disbonded FBE coating, Figure 4-6c shows that the solution pH increases more quickly, and reaches up to 10.4 after 30 days.

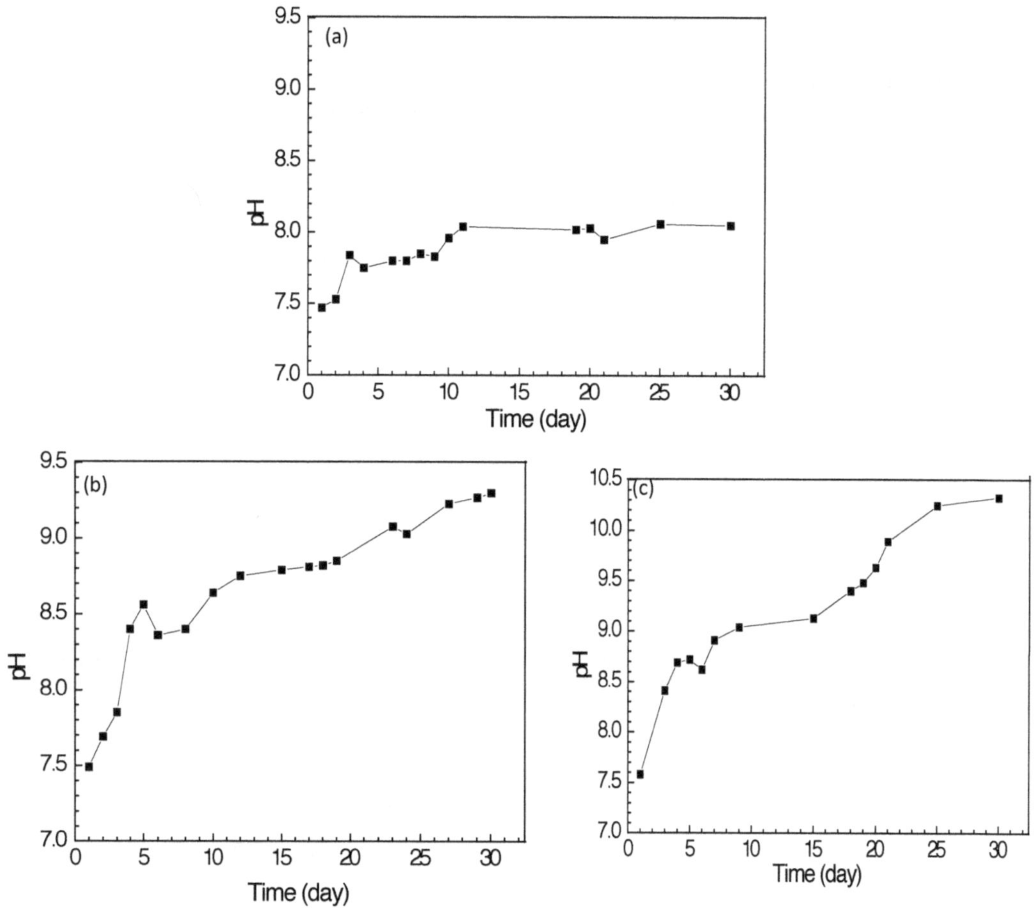

Figure 4-4. Time dependence of the solution pH under HDPE and FBE coatings at −0.85 V(CSE) and −1.00 V(CSE) CP potentials, respectively: (a) HDPE, −0.85 V(CSE); (b) FBE, −0.85 V(CSE) and (c) FBE, −1.00 V(CSE) [Kuang and Cheng, 2015a].

Figures 4-5 and 4-6 show the optical morphologies of X65 steel after various time periods in 0.01 M $NaHCO_3$ solution trapped under the disbonded HDPE and FBE coatings, respectively, at an applied CP potential of −0.85 V(CSE). It is seen that, for the disbonded HDPE, the steel experiences serious corrosion after five days. With increased testing time, more corrosion products are generated and deposited on the steel surface. Obviously, the applied CP cannot effectively protect the steel that is under a disbonded HDPE coating. For FBE coating, mild corrosion occurs on the steel electrode. Particularly, the corrosion product generated is much less than that observed in Figure 4-5, where the steel corrodes under disbonded HDPE coating. A few isolated corrosion pits form on the steel surface. The pits become deeper with increased testing time. Thus, the FBE permits CP current to penetrate and protect the steel from corrosion. However, under the given testing condition, the steel under the disbonded FBE cannot be fully protected by the applied CP, as shown by the occurrence of localized corrosion.

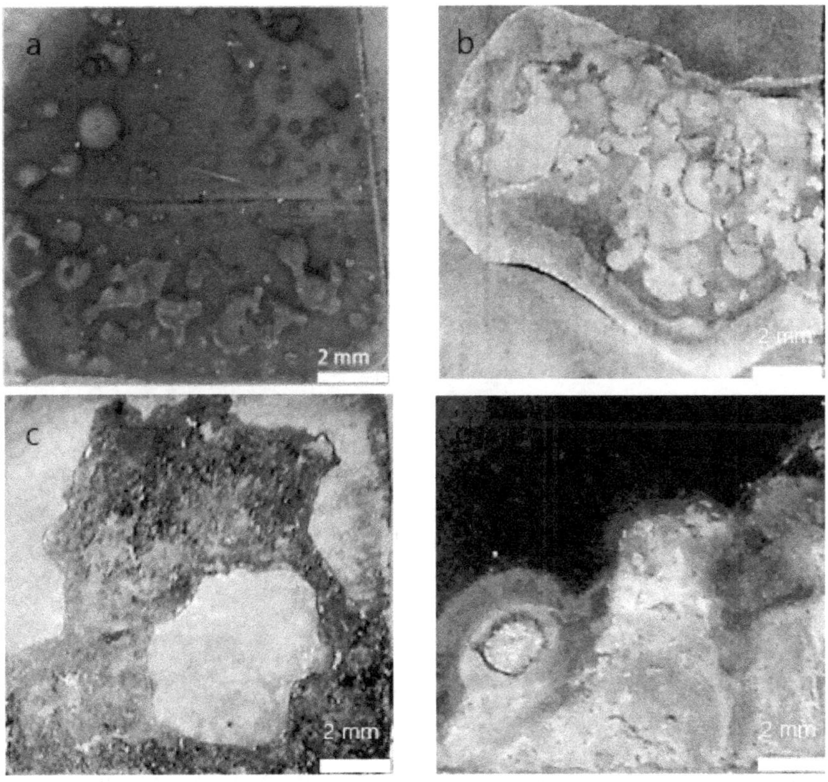

Figure 4-5. Optical morphology of the steel electrode after various time periods in 0.01 M $NaHCO_3$ solution trapped under the disbonded HDPE and FBE coatings, respectively, at an applied CP potential of −0.85 V(CSE): (a) 5 days; (b) 10 days; (c) 20 days and (d) 30 days [Kuang and Cheng, 2015a].

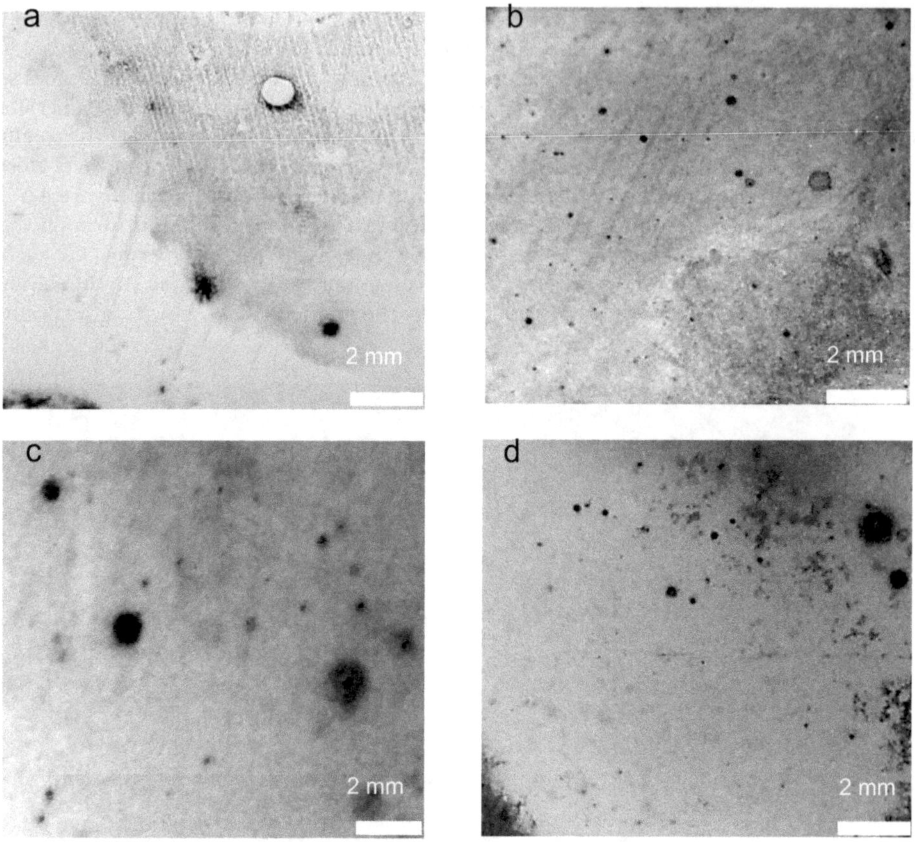

Figure 4-6. Optical morphology of the steel electrode after various time periods in 0.01 M $NaHCO_3$ solution trapped under the disbonded FBE coating at an applied CP potential of −0.85 V(CSE): (a) 5 days; (b) 10 days; (c) 20 days and (d) 30 days [Kuang and Cheng, 2015a].

From electrode potential and potentiostatic current density measurements, and from observation of the steel's corrosion morphology as well as the pH of the electrolyte trapped under disbonded coatings, the HDPE shields the CP current from reaching the steel for corrosion protection, while the FBE permits CP permeation (at least partially). Therefore, the FBE is regarded as a CP-compatible coating, and the HDPE is not. Moreover, the CP current permeation through the FBE coating film is time-dependent. With more time, more CP current permeates the coating.

What causes the distinctly different properties of the two coatings (i.e., FBE and HDPE) in terms of the CP compatibility?

The HDPE is a thermoplastic hydrocarbon material with a high molecular weight. It is widely used as a pipeline coating due to its high electric resistance and low water absorption in practice. However, research results (summarized above) demonstrate that the HDPE is a CP-shielding coating, which blocks CP current from reaching the steel for corrosion protection if the coating is disbonded and a corrosive environment is developed under the coating. Figure 4-7 shows the Fourier Transform Infrared (FTIR) spectrum of HDPE coating before and after 30 days of testing under a CP potential of −0.85 V(CSE) in 0.01 M bicarbonate solution. Find the characteristic functional groups identified in Table 4-1. The positions of the characteristic functional groups, including C–C, $C–H_2$ and $C–H_3$, are consistent with published results [Mohan and Prabakaran, 1989; Krimm et al., 1956; Gulmine et al., 2002]. Thus, there is no difference of the HDPE spectra obtained before and after testing. Moreover, there is no characteristic peak specific to the identified hydroxyl groups. This shows that water molecules are not absorbed into the HDPE. It has been acknowledged [Thomas, 1991] that water permeation into a hydrocarbon coating is predominated by its polarity. The HDPE is a non-polar polymer, and is highly resistant to water permeation. Actually, the HDPE can maintain high resistivity after a long-term water soaking, and the coating resistance is up to 10^{17} Ω cm^2 [Buchanan, 2013b]. Thus, it is the non-polar structure of HDPE polymeric molecules that makes water molecules impermeable to the coating. The high resistivity of the coating can break the CP current circuit. As a result, the HDPE shields CP from reaching the steel for corrosion protection.

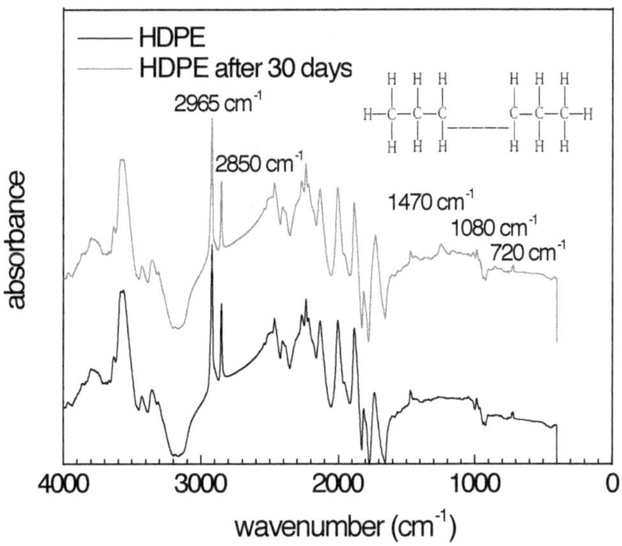

Figure 4-7. FTIR spectrum of the HDPE coating before and after 30 days of testing under a CP potential of −0.85 V(CSE) [Kuang and Cheng, 2015a].

Table 4-1. Characteristic Functional Groups of HDPE Identified in the FTIR Spectrum in Figure 4-7 [Kuang and Cheng, 2015a].

Band (cm^{-1})	Groups
2965	$C\text{-}H_3$ stretching [Mohan and Prabakaran, 1989]
2850	$C\text{-}H_2$ stretching [Krimm et al., 1956]
1470	$C\text{-}H_2$ bending [Gulmine et al., 2002]
1080	C-C stretching [Mohan and Prabakaran, 1989]
720	$C\text{-}H_2$ scissoring [Gulmine et al., 2002]

The FBE is permeable to CP, and the CP permeation is time-dependent. Within a certain time period, CP is partially permeable through the FBE coating. In other words, the applied CP potential– −0.85 V(CSE)–frequently cannot provide full protection to steel trapped under the disbonded FBE coating in corrosive environments. This finding is interesting, since FBE has been regarded as a coating with "fail-safe" characteristics in the presence of CP. Figure 4-8 shows the FTIR spectrum of an FBE coating before and after 30 days of testing under CP potential of −0.85 V(CSE). The characteristic functional groups identified are shown in Table 4-2: Observe the three characteristic absorption peaks of the oxirane ring. The peak at 831 cm^{-1} is attributed to the C–O bond of the oxirane group. The peak located at 2873 cm^{-1} is due to the C–H tension of methylene group in the epoxy ring. The next peak is located at 4623 cm^{-1}, which is a combination bond of the second overtone of epoxy ring stretching with the fundamental C–H stretching at 2873 cm^{-1} [Chike et al., 1993]. Moreover, two types of water bonds are identified in the spectra (i.e., highly mobile free water molecules at about 3600 cm^{-1}) and the combination of asymmetric stretching and bending of hydroxyl vibrations at 5215 cm^{-1} [Chike et al., 1993]. There are obvious changes of the absorbance intensities of functional groups along with time. The absorption bond of C–H stretching located at 2873 cm^{-1} becomes stronger and the bond of C–H overtone stretching of the aromatic ring at 4623 cm^{-1} becomes weaker with increasing time. Another important change is the intensity of the bonds of hydroxyl groups. The peaks of 3600 cm^{-1} and 5215 cm^{-1}, representing the absorbed water in FBE coating, become sharper with time, indicating that water uptakes occur continuously. Indeed, FBE is a highly polar coating containing hydroxyl groups. Water is also a polar molecule and capable of forming hydrogen bonds with the hydroxyl groups.

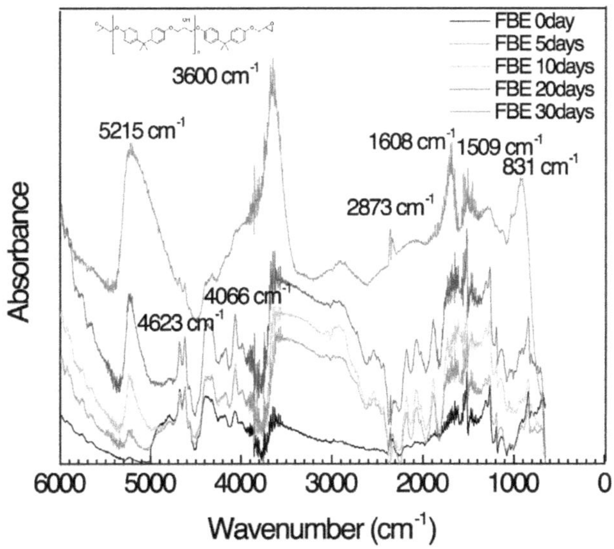

Figure 4-8. FTIR spectrum of the FBE coating before and after 30 days of testing under CP of −0.85 V(CSE) [Kuang and Cheng, 2015a].

Table 4-2. Characteristic Functional Groups of FBE Coating Identified in the FTIR Spectrum in Figure 4-8 [Kuang and Cheng, 2015a].

Bond (cm^{-1})	Groups
5215	Combination asymmetric stretching and bending of O-H [Blanco et al., 2006]
4623	Overtone of C-H stretching of the aromatic ring [Mijovic and Andjelic, 1995]
4066	Stretching C-H of aromatic ring [George et al., 1991]
3600	O-H stretching [Gonzalez et al., 2012]
2873	C-H stretching [Chike et al., 1993]
1608	Stretching C=C of aromatic rings [Mertzel and Koenig, 1986]
1509	Stretching C-C of aromatic rings [Meure et al., 2010]
831	Stretching C-O-C of oxirane group [Meure et al., 2010]

The rate of water transfer through a polymeric coating is proportional to the applied CP potential [Lee and Peppas, 1993]. The more negative the CP potential, the more cations diffuse through the coating film to decrease its resistivity. Thus, more CP current permeates through the coating for corrosion protection. Moreover, the ion-transport process inside a coating is not homogeneous, with some regions where ionic transport occurs with a high resistance and other region where ionic transport occurs with a relatively low resistance [Leidheiser and Granata, 1988]. Principally, the CP current permeates preferentially through low-resistance regions to reach the steel, causing an uneven distribution of the protective current on the steel surface. Thus, localized corrosion occurs at areas with insufficient CP. At sufficiently negative CP potentials, however–such as −1.00 V(CSE)–the enhanced electroendosmosis can drive cations to diffuse through the coating at both low-resistance and high-resistance regions, generating a uniform distribution of cathodic current on the steel for corrosion protection under the disbonded coating.

The multi-layered coatings with PE as the outer coat (such as high-performance composite coating) are usually not CP-compatible. In other words, the CP current becomes shielded from reaching the steel surface by the coating film. Figure 4-9 shows the experimental setup that enables CP permeation measurements through a coating film [Fu and Cheng, 2011], where a steel electrode and the counter and reference electrodes are installed in two chambers that are separated by the testing coating film (i.e., HPCC). An electrochemical measurement system is connected with the three electrodes for electrochemical measurements.

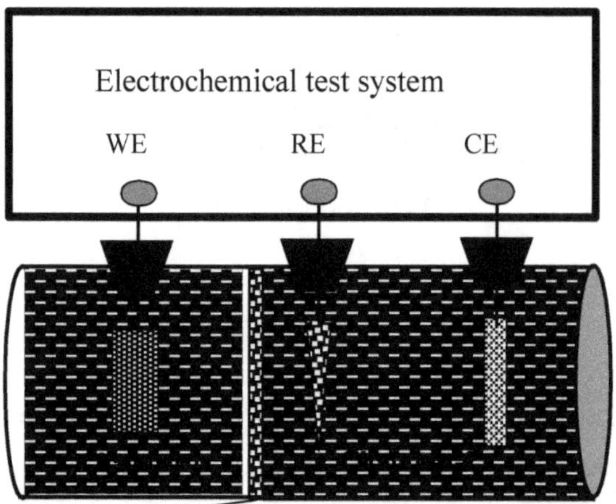

Figure 4-9. Schematic diagram of the experimental setup to study the permeability of coatings to CP through a double-chamber system [Fu and Cheng, 2011].

Figure 4-10 shows corrosion potential and potentiostatic current density measured on X65 steel that is separated from the reference and counter electrodes by a HPCC film as a function of time. The corrosion potential of the steel decreases with time. However, even after 32 days of testing, the corrosion potential is still about -0.12 V (SCE, saturated calomel electrode), which is much less negative than the corrosion potential of the steel in a bulk-soil solution of about -0.6 V (SHE, standard hydrogen electrode) [Cheng, 2007]. Moreover, at a potentiostatic polarization of −1.5 V(SCE), anodic current densities are obtained, and there is little effect from the immersion time on the measured current density. Furthermore, the electrochemical impedance spectroscopy (EIS) is measured (see Figure 4-11). For comparison, the Nyquist diagram measured on a FBE-separated double-chamber system is also present. For HPCC, the Nyquist diagrams measured at various times feature a big semicircle. Due to the random data obtained in the low-frequency range, the semicircles are incomplete. However, the Nyquist diagrams measured on the FBE-separated cell show a regular semicircle, and the size of the semicircle decreases with time.

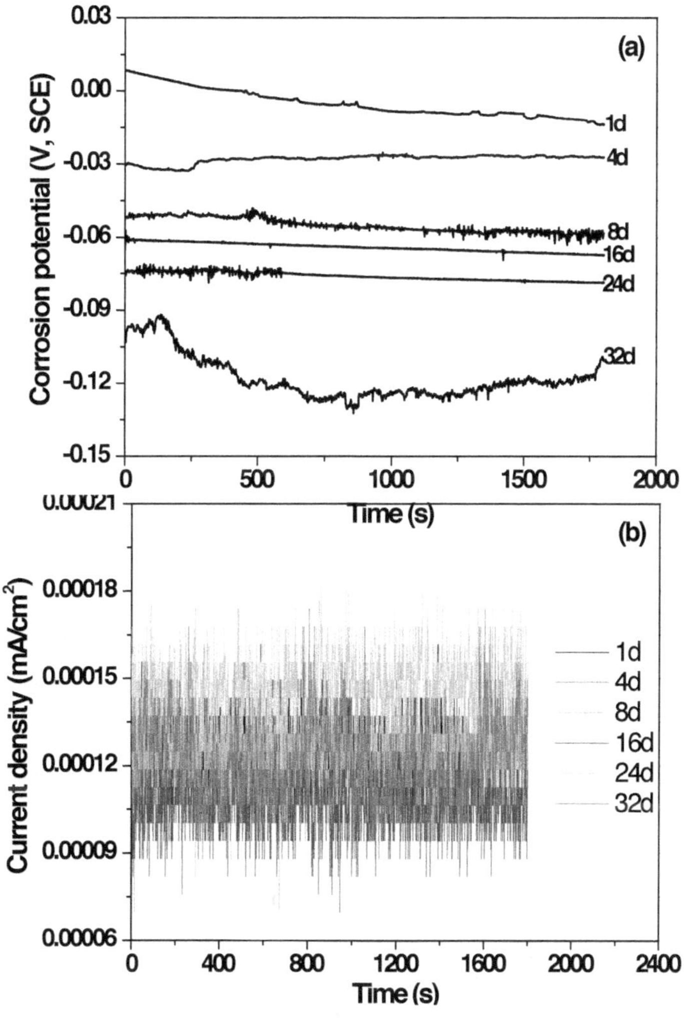

Figure 4-10. Electrochemical characterization of HPCC film as a function of time (a) corrosion potential, (b) potentiostatic current density at −1.5 V(SCE) [Fu and Cheng, 2011].

From the EIS measurements, the HPCC shows a pure capacitive behavior, with an ultra-high low-frequency impedance at the order of 10^{10} Ω cm^2 during the test period. Moreover, the potentiostatic current density measurements show that the ideal capacitive behavior of the HPCC results in anodic, rather than cathodic, current densities even when a very negative CP potential of −1.5 V(SCE) is applied. Apparently, the HPCC is impermeable to the applied CP due to its extremely high impedance. For the FBE coating, a high impedance of more than 10^7 Ω cm^2 is measured at the beginning of the

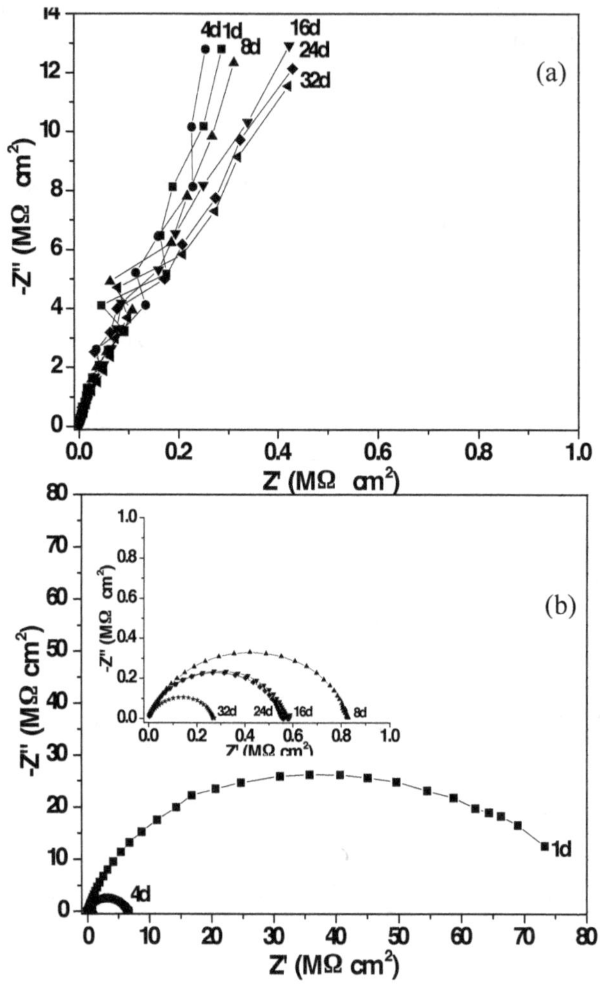

Figure 4-11. Nyquist diagrams measured on the steel in a double-cell system as shown in Figure 4-9, where the chambers are separated by (a) HPCC, and (b) FBE films, respectively [Fu and Cheng, 2011].

test. The impedance decreases with time and becomes less than 10^7 Ω cm^2 after four days of testing. When water uptake occurs, it results in decreased coating impedance. Once degraded, the coating would show a "fail-safe" feature in the presence of CP. The CP current can permeate the coating to reach the substrate steel surface for corrosion protection.

4.2.5. CP Shielding by Coating Failures—Part III. Coating Disbonding at a Holiday

Cathodic disbondment is an important mechanism that results in lost coating adhesion, which usually starts at a holiday. The applied CP at coating faults may elevate the electrolyte pH at the holiday through the enhanced cathodic reduction of dissolved oxygen or water [Perdomo and Song, 2000; Chen et al., 2009; Kuang and Cheng, 2015]. The local solution's alkalization can weaken the coating primer's bond to steel, causing coating disbondment. The disbondment, especially the bottom of the disbonding crevice, can be shielded from CP. Thus, corrosion can occur locally at the disbondment even when the pipeline is under CP.

The diagram shown in Figure 4-12 [Kuang and Cheng, 2015b] demonstrates that CP is shielded from reaching the disbonding crevice that starts at a holiday. To prepare an artificial disbondment, a coating film is applied on the surface of an X65 steel using a double-sided self-adhesive tape. The gap between the coating and the steel is defined as the disbonding thickness. The tape with a known thickness is layered to establish gaps with desired thicknesses. The coating film is then applied to the tape, which is applied to the steel surface. After that, the tape is removed to form artificial disbondments over the steel. A 10-mm diameter hole is opened on the coating to simulate a holiday, with six potential/pH micro-probes installed at distances of 30, 60, 90, 120, 150, and 180 mm from the holiday. The distance of the probing position to the open holiday is defined as the disbonding depth.

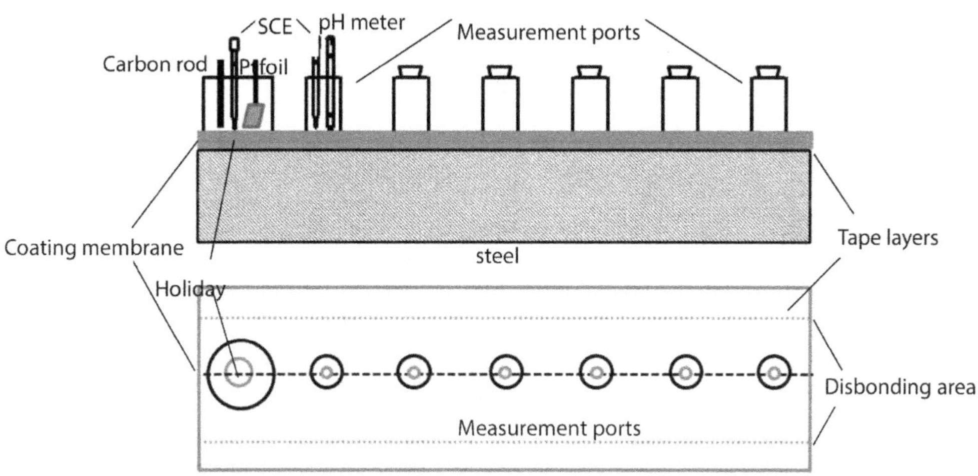

Figure 4-12. Schematic diagram of the experimental setup simulating a disbonding crevice undercoating and the local potential/solution pH measurements [Kuang and Cheng, 2015b].

Figure 4-13 shows the time dependence of the local potential distribution under a disbonded FBE coating (disbonding thickness of 120 μm) at varied disbonding depths from an open holiday, where the X65 steel was either at corrosion potential or at CP potentials of -0.875 V(SCE) and -0.975 V(SCE), respectively. The electrolyte is a 0.01 M $NaHCO_3$ solution purged with 5% CO_2 to achieve a pH of about 6.5. Prior to CP application, the local potentials at all probing positions are about -0.755 V(SCE), which is the corrosion potential of X65 steel in the solution. When the potential of -0.875 V(SCE) is applied, the potential at the holiday (i.e., 0 mm in the figure) is the applied CP value.

However, the potential at 30 mm from the holiday is less negative (i.e., -0.800 V(SCE) after 48 hours). With the increase in the disbonding depth (i.e., the probing position is farther from the holiday, the local potential is less negative than that at the 30 mm position, but the potential difference is not distinguishable). At the CP potential of -0.975 V(SCE), the potential at the holiday is still the applied value, but the local potentials at the probing positions are shifted less negatively. The disbonding depth beyond which the local potential is not distinguishable at this CP level is increased to 150 mm. Thus, the applied CP can be shielded from reaching the coating disbondment. With the increase in the disbonding depth toward the disbondment bottom, the CP shielding is more apparent. The shielding effect can be mitigated by application of more negative CP potentials.

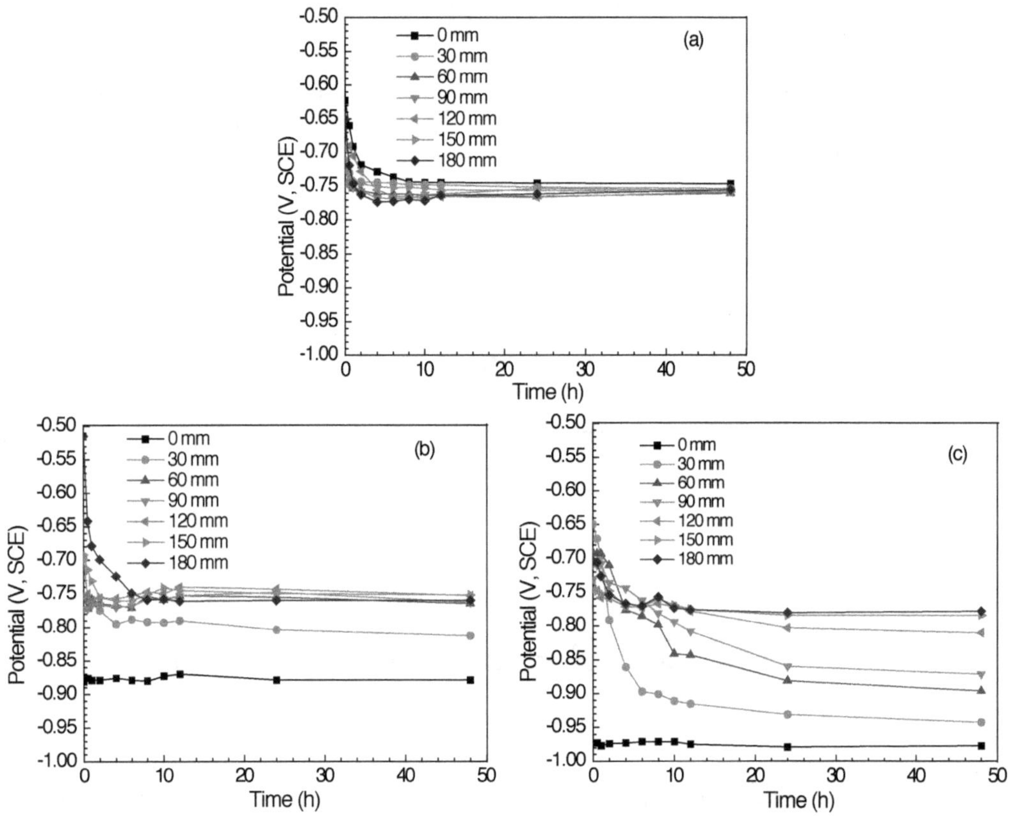

Figure 4-13. Time dependence of the distribution of local potential under disbonded FBE coating (disbonding thickness of 120 μm) at varied disbonding depths from an open holiday, where the X65 steel was either at corrosion potential or at CP potentials of -0.875 V(SCE) (a) and -0.975 V(SCE) (b), respectively. The electrolyte is 0.01 M $NaHCO_3$ solution purged with 5% CO_2 to achieve a pH of 7 [Kuang and Cheng, 2015b].

Since CP facilitates the cathodic reduction reaction to generate hydroxyl ions, raising the solution pH, the local pH can be monitored to reflect the CP potential level along the disbonding depth from the open holiday. Figure 4-14 shows the time dependence of the distribution of local solution pH under disbonded coating at various disbonding depths. Prior to CP application, the solution pH is about 7.5 (i.e., the value of the prepared solution) at all probing positions. Upon CP application, the solution pH is elevated. Moreover, when the CP potential is more negative, the solution pH is further elevated at individual probing positions. For example, at the open holiday, the steady-state solution pH is 8.5 at -0.875 V(SCE) and 11.0 at -0.975 V(SCE). However, the CP-driven pH elevation becomes less obvious with the increasing disbonding depth, especially at the disbondment bottom due to the CP shielding.

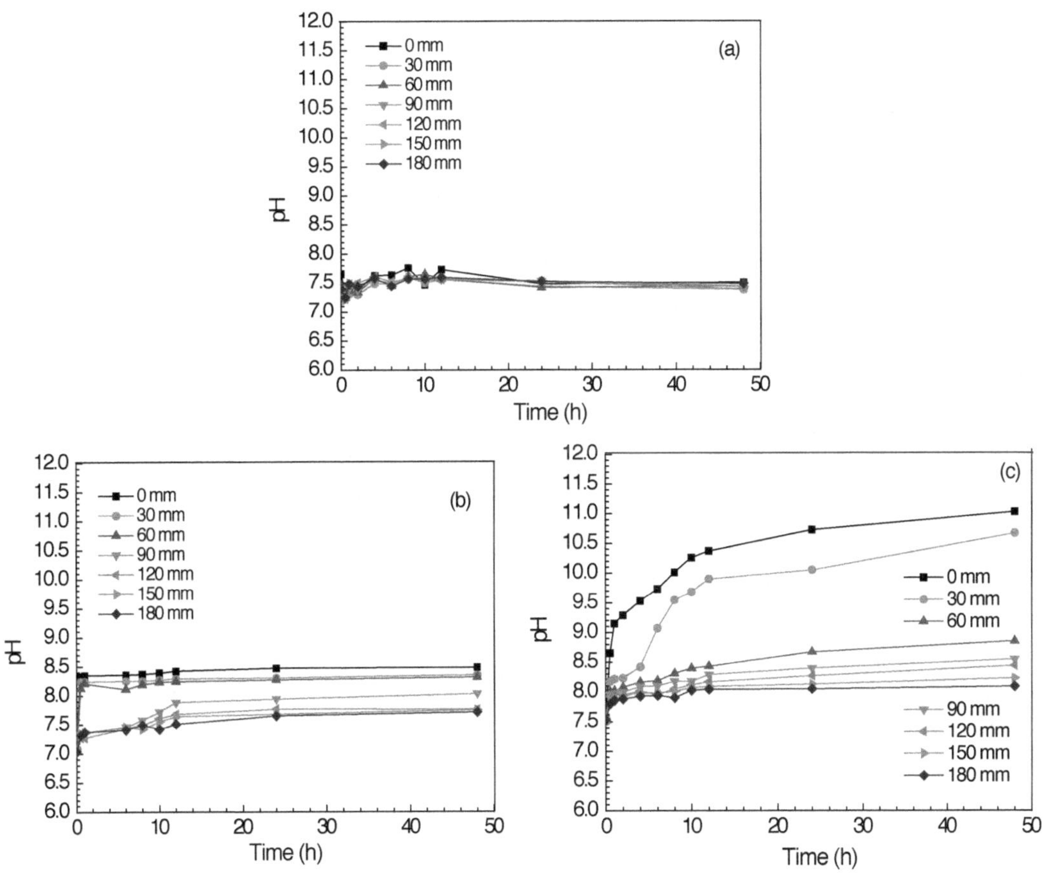

Figure 4-14. Time dependence of the distribution of local solution pH under disbonded coating at various disbonding depths under the testing conditions same as those in Figure 4-13 [Kuang and Cheng, 2015b].

When CP is applied to coated steel where holidays are contained in the coating, the CP is primarily applied to the open holiday. The solution pH at the holiday is elevated with the negative shift of the CP potential. The non-uniform distribution of solution pH from the open holiday to the disbondment indicates that the CP-induced pH elevation is not fully realized under the disbonded coating. The measurements of the potential distributions from the holiday toward the disbondment bottom indicate that the applied CP is shielded (at least partially) under the disbonded coating. At the disbonding thickness of 120 μm, only the open holiday is under the full CP potential. With increasing distance from the holiday (i.e., the increasing disbonding depth) the local potential tends to be less negative until the steady-state corrosion potential is reached. Due to the shielding effect, the steel under the disbonded coating is not under CP. Thus, corrosion occurs. The potential records also indicate that, to cathodically protect the steel under coating disbondment, the CP potential must be sufficiently negative. Hydrogen evolution must also be considered when the CP potential is too negative.

The shielding effect of coating disbondment on CP permeation is affected by the disbonding thickness. Figure 4-15 shows the distribution of local potential under a disbonded FBE coating at various disbonding depths from the open holiday, where the CP potential of -0.875 V(SCE) is applied under various disbonding thicknesses. Identical to previous results, the CP is shielded from reaching the disbondment. Only at the open holiday is the measured value the same as the applied CP potential. Under the coating disbondment, the potential tends to be less negative. Moreover, with an increase in disbonding thickness, the CP shielding effect becomes less significant. For example, at the disbonding thickness of 120 μm, the local potential at the probing position of 30 mm is about -0.810 V(SCE). When the disbonding thickness increases to 240 and 360 μm, the potentials at the same position are -0.855 and -0.865 V(SCE), respectively. Therefore, as the coating disbondment becomes wider (i.e., with an increased disbonding thickness), the CP shielding effect is less significant.

Figure 4-16 shows the distribution of solution pH under a disbonded FBE coating at various disbonding depths from the open holiday, where a CP potential of –0.875 V(SCE) is applied under various disbonding thicknesses. The applied CP can elevate the solution pH, especially at the open holiday. With the increasing disbonding depth, the solution pH tends to be the value of the originally prepared solution. As the disbonding thickness increases, the CP-induced pH elevation becomes more obvious, even at the disbondment bottom. For example, at the disbonding thickness of 120 μm, the solution pH at the disbondment bottom, i.e., 180 mm from the holiday, is only about 7.5. This indicates that the CP does not penetrate into the disbondment bottom to raise the local pH. When the disbonding thickness is increased to 240 μm and 360 μm, the solution pH values at the disbondment bottom are 7.9 and 8.1, respectively. Thus, as the coating disbondment widens, the CP-enhanced pH elevation is more appreciable.

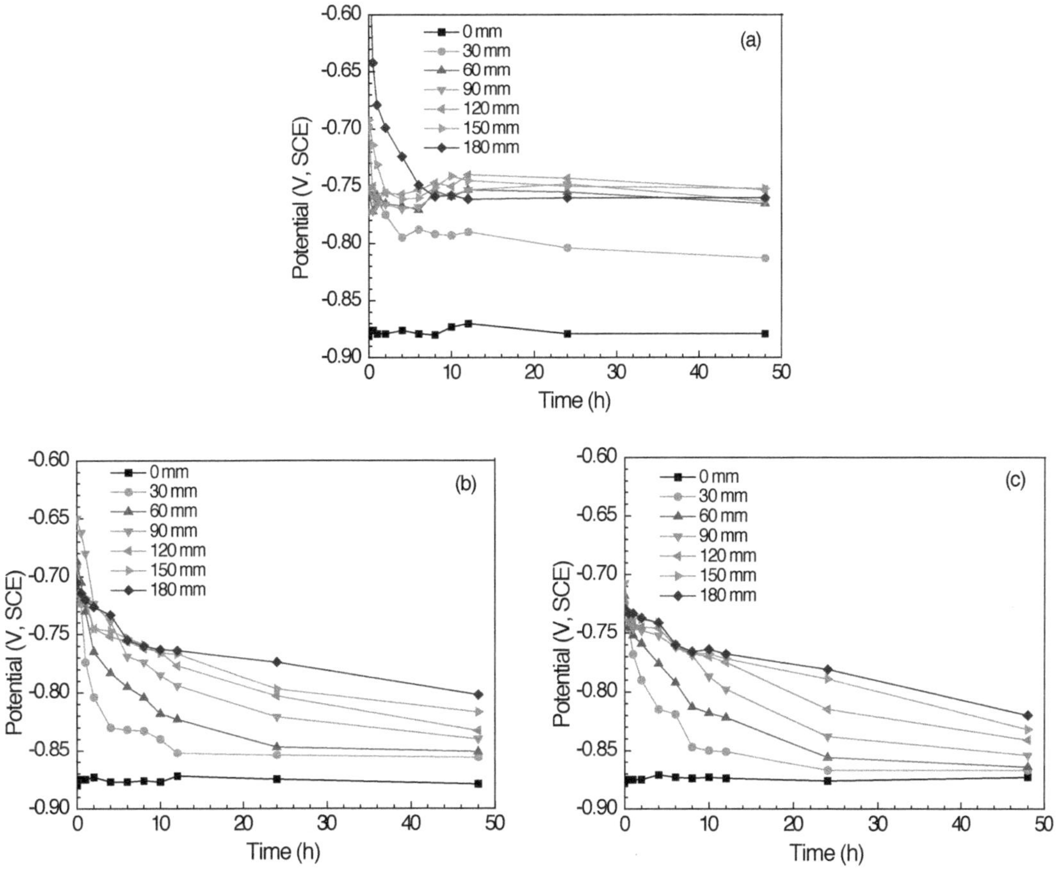

Figure 4-15. Distribution of local potential under a disbonded FBE coating at various disbonding depths from the open holiday, where the CP potential of -0.875 V(SCE) is applied under various disbonding thicknesses (a) 120 μm, (b) 240 μm, (c) 360 μm [Kuang and Cheng, 2015b].

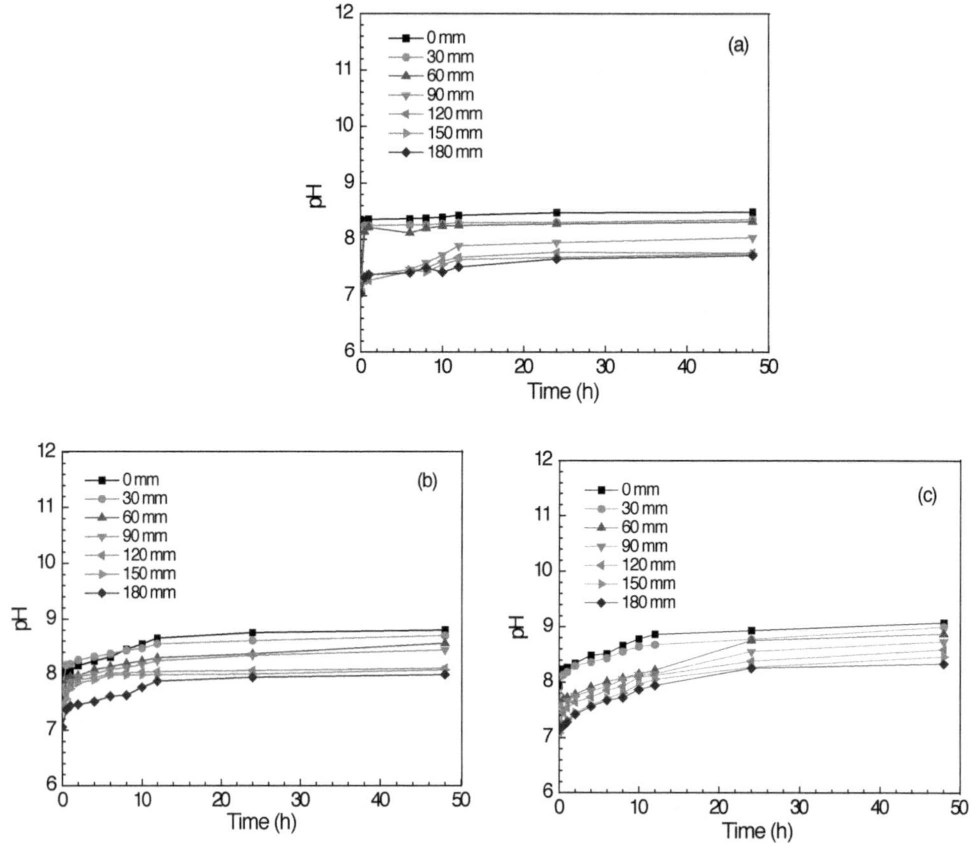

Figure 4-16. Distributions of solution pH under disbonded FBE coating at varied disbonding depths from the open holiday where the CP potential of -0.875 V(SCE) is applied under various disbonding thicknesses (a) 120 μm, (b) 240 μm, (c) 360 μm [Kuang and Cheng, 2015b].

The measurements of both local potential and solution pH under a disbonded coating show that the CP shielding tends to be mitigated when the disbonding thickness is increased, and the potentials under the disbonded coating approach those at the open holiday. Moreover, the pH elevation under disbondment is more apparent. Thus, the geometrical factor of the coating disbondment plays an essential role in CP shielding.

CP shielding by coating disbondment can result in cathodic polarization of pipeline steel at the holiday, while the steel at the disbondment bottom is still at its corrosion potential, depending on the disbonding thickness and applied CP. The potential difference causes separated anode and cathode sites, where the cathodic reaction occurs at the holiday and the anodic reaction at the disbondment bottom. The disbondment can be full of corrosion products, which are difficult to diffuse from the disbonding crevice due to limited geometrical space. This further increases the blocking effect on CP permeation. This is the key mechanism of localized corrosion on pipelines that are under CP. This phenomenon is demonstrated by frequent field experiences in which extensive corrosion pits are found under disbonded coating on cathodically protected pipelines [Baker, 2004].

4.2.6. CP Shielding by Coating Failures—Part IV. Effect of Alternating Current Interference

Buried pipelines corrode at an accelerated rate in the presence of alternating current (AC) interference from adjacent high-voltage power transmission lines [Fu and Cheng, 2010; Fu and Cheng, 2012; Kuang and Cheng, 2014]. In addition to enhanced external corrosion and the shift of CP potential from the applied value [Xu and Cheng, 2013], the AC can result in coating disbondment and affect CP shielding beneath it [Kuang and Cheng, 2015c].

A test setup similar to the one shown in Figure 4-12 is used to study the shielding effect of coating disbondment when the coated steel is under various AC current densities. Figure 4-17 shows the potential distribution under disbonded FBE coating in the trapped 0.01 M $NaHCO_3$ solution (disbonding thickness of 120 μm) at CP potential of -0.875 V(SCE) under various AC current densities. In the absence of AC, the CP is shielded to reach the disbonding crevice along the depth direction, which is consistent with previous measurements. Particularly, the critical depth in the absence of AC is 60 mm, above which CP cannot reach, and the steel is at free corrosion potential of -0.755 V(SCE). Upon AC application, the DC potential of the steel shifts negatively. Moreover, as AC current density increases, the potential shift is more negative as seen from the potential measurements on the holiday (distance of 0 mm). Moreover, along the direction towards the disbonding depth, the potential is shifted less negatively. CP shielding due to the crevice geometry also exists under AC interference. It is also interesting to note that, at small AC current densities (such as 100 A/m^2), the critical crevice depth where the CP cannot reach is about 150 mm from the open holiday. With an increasing AC current density, the critical crevice depth decreases. When AC current density is up to 200, 300, and 400 A/m^2, the critical depths are 120, 90, and 30 mm, respectively. When the AC current density is 500 A/m^2, there is no CP that can permeate the potential monitoring regions. Thus, at small AC current densities, the AC results in more negative potentials at individual disbonding depths compared to those measured in the absence of AC. However, the trend is diminished with increasing AC current densities. When comparing potential profiles measured in the absence of AC (Figure 4-17a) and at an AC current density of 400 A/m^2 (Figure 4-17e), there is a high similarity under the disbonding crevice, except that the latter has a more negative potential at the open holiday. At a large AC current density such as 500 A/m^2, the CP is completely shielded from reaching the disbonding crevice.

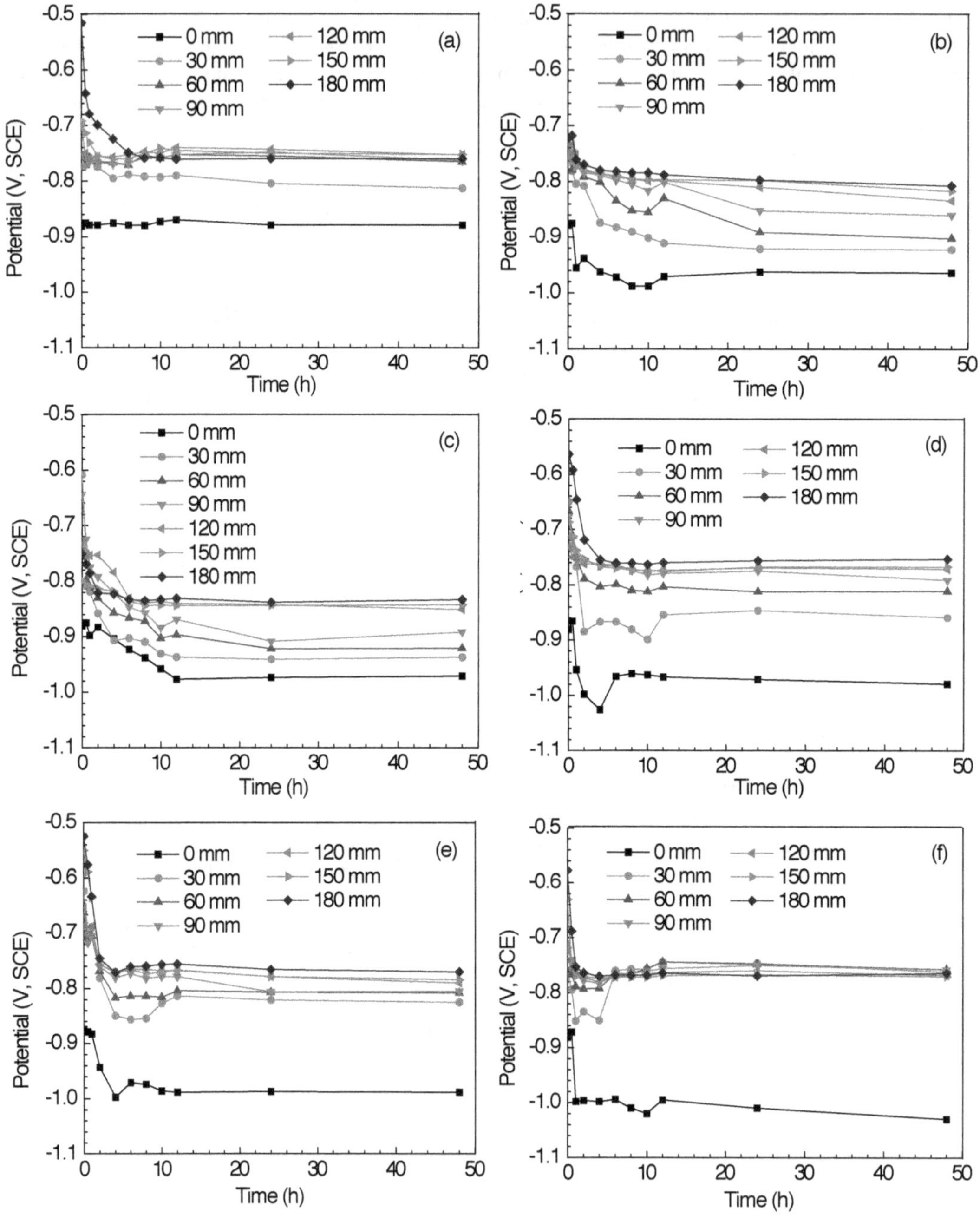

Figure 4-17. Potential distribution under disbonded FBE coating in the trapped 0.01 M $NaHCO_3$ solution (disbonding thickness of 120 μm) at CP potential of -0.875 V(SCE) under various AC current densities (a) 0 A/m^2, (b) 100 A/m^2, (c) 200 A/m^2, (d) 300 A/m^2, (e) 400 A/m^2, (f) 500 A/m^2 [Kuang and Cheng, 2015c].

Therefore, AC affects the CP current permeation into the electrolyte under disbonded coating on pipelines. The AC application negatively shifts the steel potential at the open holiday. This is attributed to accelerated diffusion of cation ions that are generated during steel corrosion under an AC-enhanced electric field [Kuang and Cheng, 2014; Xu and Cheng, 2013]. As a consequence, the number of positive charges in the double-charge layer decreases, resulting in a negative shift of the potential. To clearly show the effect of AC current density on CP permeation into the disbonding crevice, the local potentials at the disbonding sites of 30 and 180 mm from the holiday are plotted as a function of AC current density, as shown in Figure 4-18. At small AC current densities such as 100 and 200 A/m^2, the local potential at both sites is shifted negatively by AC. As discussed, this shift is associated with the enhanced corrosion of steel under AC interference. When the AC current densities increase from 300 to 500 A/m^2, the potentials are shifted towards the positive direction. At high AC current densities, the iron dissolution and water reduction are further enhanced in the positive and negative cycles of the AC, respectively. There are more ferrous and hydroxyl ions generated during anodic and cathodic reactions. Due to the limited space under the disbonding crevice, corrosion products cannot diffuse outwards freely. The positive shift of the steel potential is associated with a deposit of corrosion product, such as $Fe(OH)_2$, on the steel.

Furthermore, at small AC current densities, the AC-enhanced steel corrosion generates ferrous and hydroxyl ions, improving the conductivity of the solution trapped under the disbonded coating. This results in an enhanced permeation of CP current into the crevice, relieving the CP-shielding effect. With an increase of the AC current density, the critical disbonding depth decreases where the CP current can reach, which is associated with the blocking effect of corrosion products formed under the coating. The corrosion products can block the ionic diffusion and CP current, thus decreasing CP permeation. When the AC current density is increased to 500 A/m^2, the CP is completely blocked from reaching the disbonding crevice. It is important to note that, under high AC current densities, the applied CP can be shielded completely from reaching the coating disbondment.

CP shielding can also occur at coating defects, where the defect is featured with a deep, narrow geometry. In other words, the aspect ratio of the coating defect (i.e., the ratio of the defect depth to its width or diameter), is critical to the CP current permeation into the bottom of the defect for corrosion protection. Details about CP shielding at coating defects will be further discussed in Chapter Seven.

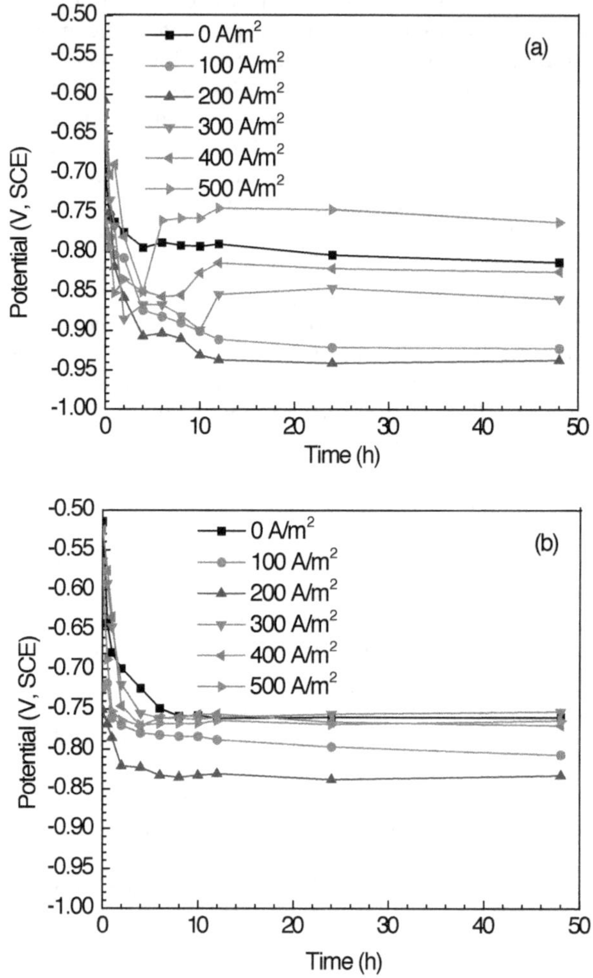

Figure 4-18. Local potentials at disbonding sites of (a) 30 and (b) 180 mm from holiday as a function of AC current density (CP potential: -0.875 V(SCE), disbonding thickness: 120 μm) [Kuang and Cheng, 2015c].

4.3. Failure and Effect Analysis for Impermeable Coatings

Permeability of coatings is defined as a measure of the ability of the coating to be permeated or penetrated by liquid, gas, or vapor. High permeability means the permeate (permeating substance) passes easily through the coating film. The permeation typically involves both dissolution and diffusion. For instance, water vapor permeates a coating by getting adsorbed at the entry face, dissolving a part of the film, reaching equilibrium, and then follows the diffusion through the film and finally desorbs at the other side of the film. The dissolution involves the action of chemicals on the polymers (usually in waterborne coatings), polyethylene oxides, resins of silicone, surfactants, and other residual water-soluble solvents. Diffusion can be affected by pores, cracks, and other defects present in the coating. Even if the coating film is free of defects, diffusion can still occur. The concentration and nature of pigments have considerable effects on diffusion and, hence, the permeability of coatings

4.3.1. Characteristics of Impermeable Coatings

To serve as an effective physical barrier for corrosion protection between the underlying steel and the external environments, a good pipeline coating should limit the penetration and diffusion of water, oxygen, carbon dioxide, and ionic species across it. A coating system's impermeability refers to its ability to block the passage of water, gases, and chemicals across the coating to reach the steel surface. Actually, no polymeric coating is completely impermeable, but the polymeric coatings vary in their impermeability properties. As a result, the definition of a coating's "impermeability" is relative and does not have a quantitative threshold for "high" or "low" permeability or impermeability. In other words, all polymer-based coatings for pipeline use have permeable and impermeable properties.

An impermeable coating is often used for structures that are serviced in immersion environments, and must be inert to surrounding chemicals. It must also be impervious to air, oxygen, carbon dioxide, and the passage of chemical ions. It is dielectric and breaks the electric circuit between coated metal and the environment, which is usually an electrolyte when wetted. In summary, an impermeable coating forms an inert barrier over the surface of the structure to be protected.

The permeability property of a coating depends on factors such as a coating's chemical nature and structure, its thickness, and the method of measurement [Mayne, 1952]. As a typical example, Table 4-3 shows the permeation coefficient (mol m/m^2 s atm) for O_2, H_2O, and CO_2 for FBE, ethylene-vinyl acetate (EVA), HDPE and polyvinylidene chloride (PVDC) coatings, as used in Fick's first law to determine the permeation rate in the coating by each of these corrosive species under different relative humidities [Song et al., 2005]. Coatings possess different permeabilities to the species.

Table 4-3. Permeability Coefficient of Various Coatings to O_2, H_2O, and CO_2 [Song et al., 2005].

Coatings	50% RH, O_2 ($\times 10^{13}$)	90% RH, H_2O ($\times 10^{10}$)	50% RH, CO_2 ($\times 10^{13}$)
FBE	20	6	16
EVA	1550	9	5683
HDPE	454	1	517
PVDC	3	1	10

HDPE is highly impermeable to the passage of water and other aqueous chemical solutions [Fujita, 1968], but compared to epoxy coatings such as FBE, it has a higher permeability to gases such as O_2 and CO_2. For the purpose of this book, impermeability is defined in terms of the coating's ability to block water and aqueous species, which also include dissolved O_2 and dissolved CO_2. Permeation of these species through a coating is directly associated with a CP current permeating the coating, a key characteristic a pipeline coating should own, as stated above.

A typical example of impermeable coatings, based on the definition mentioned above, is a PE-based coating. Any coating that contains a PE layer can be considered an impermeable coating. The failure mode and effect analysis (FMEA) of PE-based coatings generally apply to that of impermeable coatings.

Impermeable coatings, including PE, possess the desired characteristics described below.

High resistivity. To be corrosion-resistant, a coating must break the electric circuit that causes corrosion reaction. The PE has a dry-volume resistivity of about 10^{17} Ω cm that is 10–100 times the magnitude higher than FBE of approximately 10^{15} Ω cm. Generally, a resistivity greater than about 10^{9} Ω cm is considered non-conductive [Buchanan, 2013a]. While the high resistivity of a coating stops corrosion, it can also block CP current to reach the underlying metal if the coating is disbonded.

Impermeability to water. Entry of water into a coating is the first step in the development of corrosion environments on a metal surface under disbonded coating. Thus, the impermeability to water is an important property of impermeable coatings for corrosion resistance. It was reported [Buchanan, 2013a] that the absorption of water by PE is less by a magnitude of 40 times compared to FBE.

Excellent mechanical wear resistance. Various test results show that PE is very abrasion-resistant, even compared to concrete and metals. It was demonstrated [Gabriel and Moran, 1998] that the PE has a wear rate up to 10 times less than steel. Among a number of coating materials, PE offers the maximum resistance against mechanical wear.

Excellent chemical resistance. The stability of coatings in service environments is critical for their performance and long-term applicability. PE is widely recognized for its unique chemical resistance to a wide variety of chemicals. Since the PE is virtually immune to electrolytic attack, aqueous solutions of chemicals, acids, and bases have no adverse effect. Actually, nearly all manufacturers list the barrier of HDPE to water, chemicals, acids, and bases as excellent [IPEX, 2009].

Impermeable coatings also have some undesired properties, including the following:

Lack of adhesion. The most important limitation of impermeable coatings is a lack of sufficient adhesion to the substrate metal, such as pipeline steels. Generally, PE has little tendency to bond with metals. Thus, in dual and three-layer coating systems (such as Yellow Jacket and 3LPE coatings), PE is mainly used as the outer layer over a specially modified adhesive when a primer with a strong bonding to the substrate (such as FBE) is used. For example, in 3LPE, the adhesive and PE outer layers are extruded over an FBE primer.

Permeability to hydrocarbons. Another undesirable characteristic of PE as a coating is its permeability to hydrocarbons. This is a potential problem that causes PE degradation especially when it is used as a liner for hydrocarbon-transporting pipelines. Moreover, PE is a non-polar material and is permeable to non-polar molecules, including O_2, CO_2, methane [Ethridge et al., 2008], and a large number of petroleum hydrocarbons such as aliphatic, aromatic, and asphaltic [Ritums et al., 2006]. The permeation is usually enhanced at elevated temperatures and pressures.

CP shielding. As stated, impermeable coatings (such as PE and PE-based coatings) are effective in blocking water and aqueous species that may permeate the coating. This, at the same time, breaks the CP current flow circuit, disabling the CP performance. The CP shielding is one of the most common mechanisms resulting in corrosion and SCC under disbonded coating on oil/gas pipelines.

4.3.2. Coating Disbondment

As stated, impermeable coatings such as PE possess a weak bonding strength to the substrate metal. Frequently, field observations have indicated that corrosion occurs under disbonded, impermeable coatings, especially at the bottom of the disbondment. The electrolyte can be generated due to permeation of water, chemicals, CO_2, etc. from seams (for wrapped tape coating) or local openings (usually at pipeline welds) into the disbondment. The electrolyte can be a layer of solution or even water condensate with a thickness of tens of microns. As CP is shielded from reaching the disbondment by the impermeable coating, corrosion occurs on the steel.

The electrolyte generated under a disbonded impermeable coating (such as PE tape) is usually a diluted, near-neutral pH bicarbonate solution. The solution pH is in the range of 5.5–7.0, and the bicarbonate concentration is about 0.01 M. When the coating is intact (i.e., there is no defect such as pinhole or holiday contained in the coating), the electrolyte trapped under the coating disbondment is anaerobic with a trace amount of oxygen. Moreover, the applied CP is shielded by the coating. As a result, the electrochemical potential of the pipe steel is at its corrosion potential, which is about -750 mV(CSE). Both field experiences and lab tests demonstrate that this environment could result in a near-neutral pH SCC on pipelines, where both anodic dissolution reaction and the hydrogen effect are involved in the crack propagation. Details about the coating failure and the resulting near-neutral pH SCC will be described in Chapter Five.

If an impermeable coating is disbonded locally due to poor surface preparations or coating application procedures, a blister and blisters would form over a certain area. Corrosion does not occur on steel under the blister if a corrosive environment does not generate. When the blister is opened, the steel is usually dry and there is no corrosion sign.

4.3.3. Pinholes and Holidays

While pipelines are coated infrastructure and CP is applied for corrosion protection, external corrosion can occur if there is a coating discontinuity, where the pipe steel is in contact with the soil, which is usually an effective electrolyte to cause steel corrosion. Almost all polymeric coatings have inherent, unavoidable defects (e.g., pinholes, holidays, etc.), regardless of their formulations or application processes. These defects serve as channels for diffusion of water, moisture, and chemical species to reach the steel surface. Therefore, corrosion could occur locally, where coating adhesion is lost. The other effects due to the presence of pinholes and holidays include imbalanced conductivity and dielectric strength between the defected area and other intact areas.

Generally, the coating disbondment frequently starts at a holiday or pinhole. The CP-driven electrochemical reduction of water (the reduction of dissolved oxygen is negligible due to the anaerobic environment developed under impermeable coatings) generates hydroxyl ions, elevating pH of the solution. A local alkalinity is thus generated to facilitate disbonding of the coating from the holiday or pinhole to a certain extent, resulting in cathodic disbondment. Since CP can be shielded from reaching the disbondment, especially its bottom, corrosion would occur locally, resulting in the generation of corrosion pits although the pipeline is under CP macroscopically.

Localized corrosion can also occur at the bottom of the defect (such as with a pinhole or holiday) when CP current is shielded, either completed or partially, from reaching the defect base. Generally, the CP performance at coating defects depends on defect size, especially their aspect ratio (i.e., the ratio of the defect width to its depth). For big defects present on thin coatings, the aspect ratio is sufficiently large to allow CP to reach the defect bottom for corrosion protection. However, for small defects on thick coatings, the CP becomes shielded, at least partially, from reaching the base steel. The CP shielding at coating defects will be detailed in Chapter Seven, where the localized electrochemical impedance spectroscopy technique will be introduced for characterization of corrosion behavior at these defects.

In practice, industry records the average (or macroscopic) CP data only. The actual CP potential at coating defects, especially at the defect bottom, is difficult to measure, if not impossible, and thus ignored. Even if the recorded CP potentials meet the industry standard, localized corrosion (pitting corrosion) can still occur at the defect base on pipelines. Furthermore, the CP shielding at coating defects depends on the defect geometry and is independent of the type of the coating (i.e., impermeable or permeable coating).

4.3.4. Missing Coating

Coating can be extensively damaged and missing. In other words, the pipe is uncoated over a certain area. The CP can reach the steel for corrosion protection without the shielding effect mentioned above. From the viewpoint of corrosion protection, the missing coating is not a critical failure mode. However, the CP current demand is increased. Obviously, missing coating is not economically viable. Immediate actions should be taken to repair (recoat) if missing coating is found. Whether the coating is impermeable or permeable does not matter in terms of corrosion protection by CP when the coating is missed over a certain area.

4.3.5. Permeability of the Coating

Despite the high chemical and water resistance of impermeable coatings, they are permeable to gases such as O_2 and CO_2. When gases permeate the coating, which is well-bonded to substrate steel, the pressure within the coating increases. At high levels of gas permeation, the build-up of pressure may be reduced by liberation of gases, which usually occurs at the coating/steel interface and results in disbondment of the coating. At low levels of permeation, the state of equilibrium can be reached without any chemical or physical changes to the coating.

Some grades of LDPE coating can show excessive permeability to water vapor. In this situation, the coating has lowered adhesion (Norman & Argent, 2007). Moreover, the conductivity of the coating increases and its dielectric strength decreases.

In summary, as the impermeable coatings (such as PE or PE containing multi-layered coatings) are highly resistant to permeation of water and chemicals, appreciable penetration of these species will happen only after a long period of service. Therefore, this is not a major mechanism resulting in degradation for impermeable coatings.

4.4. Failure and Effect Analysis for Permeable Coatings

4.4.1. Characteristics of Permeable Coatings

A permeable coating refers to a coating that allows the penetration of water and chemicals through the coating film. Excellent permeable coatings also possess properties such as a high resistivity, good dielectric strength, excellent adhesion, high impact and mechanical strengths, etc. Different from impermeable coatings, the permeable coating permit permeation of water, moisture, chemicals, etc., into the coating, especially after long-term service in the field. Moreover, one of the most important characteristics of a permeable coating is its compatibility with CP, i.e., the ability of the coating to allow CP current to penetrate through it. Thus, this type of coating is named as "non-shielding" or "CP compatible" coating. When a permeable coating degrades, the pipeline is still protected from corrosion as the CP current can pass through the coating. The permeable coating is also named the so-called "fail-safe" coating. A typical example of permeable coatings is FBE coating. When selecting a pipeline coating, the "Fail-Safe" (or non-shielding) characteristics may be more important than other issues that are normally considered. [Norsworthy, 2006]

The FBE is highly adhesive and flexible, provides a good resistance to chemicals and, very importantly, allows CP current to penetrate. If the coating is disbonded and a corrosive electrolyte is trapped beneath it, the CP current can pass through the coating to reach the pipe steel for corrosion protection. The major limitation of permeable coatings such as FBE includes a low impact resistance. As the FBE is a plant-applied coating, it can be damaged during storage, transportation, and pipeline construction. Currently, double-layer FEB coatings have been developed and applied on pipelines to improve their resistance to mechanical damage [Pratt et al., 2011].

4.4.2. Coating Disbondment

Permeable coatings are compatible with CP. Normally, the permeation of CP current protects the steel pipe from corrosion. However, due to the cathodic reduction of water or dissolved oxygen by CP-driven electrochemical reductive reactions to generate hydroxyl ions, the pH of the trapped water is elevated. As a result, a local alkaline environment is generated under the coating, leading to the coating's cathodic disbonding from steel substrate. Moreover, the permeable coating can become disbonded locally due to poor surface preparations or coating applications. As a result, blisters would be formed. Since the coating is permeable to water and chemicals, corrosive environments are expected to generate under the blister. However, the coating is permeable to CP current, which protects the pipe steel from corrosion even when the corrosive electrolyte is generated under the coating.

While coating disbondment is a more critical issue for impermeable coatings that shield CP from reaching the steel surface for corrosion protection, the cathodic disbonding of permeable coatings is also a serious threat to pipeline integrity. In reality, the CP can become somewhat shielded regardless of whether the coating is permeable or not. Corrosion products such as $Fe(OH)_2$ are generated in the alkaline environment that is generated due to CP-driven electrochemical cathodic reactions. The corrosion products deposit in the disbondment due to limited diffusion in the crevice space, blocking the penetration of CP current into the bottom of the disbondment. Therefore, even for the permeable coating such as FBE, the CP shielding under coating disbondment is an issue to pipeline integrity, as recorded in Figures 4-13 and 4-15.

4.4.3. Pinholes and Holidays

Identical to the effect discussed with impermeable coatings, if an effective CP is available and reaches the bottom of the pinhole or holiday, localized corrosion can be prevented. However, the CP can become shielded from reaching the defect bottom, depending on the geometry of the defect. Thus, corrosion can potentially occur at the base of the defect, no matter what the permeability of the coating is.

4.4.4. Missing Coating

As stated for impermeable coatings, a missing coating does not present a critical failure mode because CP can provide sufficient protection to the exposed pipeline steel. However, it will result in a significant increase in demanded CP current. Actually, the effect of a missing coating on pipeline integrity is identical for both impermeable and permeable coatings.

4.3.5. Permeability of the Coating

The permeation of permeable coatings to environmental species, such as water, moisture, O_2, CO_2, chemical ions, etc., can degrade the coating properties and performance. There is an increase in conductivity of the coating with time, reducing the coating resistance. Moreover, the movement of moisture and ions can lead to the loss of the coating adhesion, which eventually causes disbonding. Decreased dielectric strength is another direct effect of the permeation of environmental species. The coating's dielectric strength decreases as it ages. For permeable coatings, permeability is an important parameter to evaluate and predict the long-term field performance of the coating, and thus, a further discussion about the coating permeability is an important integrant to the coating properties.

Permeability, *P*, of a material such as coating film is defined as the amount of a permeant transporting through the film in unit area and unit time by unit driving force gradient (usually partial pressure) [Jost, 1961], i.e., *P* = (Amount of permeant transported) /(Area of coating film × Time × Driving force gradient), which can be expressed as Eq. (4-5),

$$P = \frac{Q}{A\left(\frac{p_1 - p_2}{L}\right)} \tag{4-5}$$

where Q is the amount of the permeant permeated through a coating film, A is the cross-sectional area of the coating film, p_1 and p_2 are partial pressures of the permeant (if it is a gas) at both sides of the coating, respectively, and L is the coating thickness.

According to mass balance, the mass flow, Q, of the permeant transporting through the coating is equal to the change of the amount of the permeant in the detection chamber, which is written as:

$$Q = V\frac{\Delta c}{\Delta t} \approx V\frac{dc_2}{dt} \tag{4-6}$$

where V is the volume of the detection chamber, t is time, and c_2 is the concentration of the permeant in the detection side of the coating.

The rate of transfer of the permeant can also be written as:

$$Q = \frac{PA(p_1 - p_2)}{L} = \frac{PA(c_1/S - c_2/S_w)}{L} \tag{4-7}$$

where S_w is the solubility of the permeant, such as a gas, in water (0.032 mol/L atm for CO_2 and 0.0014 mol/L atm for O_2), and c_1 is the concentration of the permeant in the charging side of the coating. Combine Eqs. (4-6) and (4-7):

$$V\frac{dc_2}{dt} = \frac{PA(c_1 - c_2)}{S_w L} \tag{4-8}$$

Since c_1 (i.e., the concentration of the permeant in the charging side) can remain constant, Eq. (4-8) is rewritten as:

$$Q = \frac{VLS_w}{A}\frac{dc_2}{(c_1 - c_2)dt} = \frac{VLS_w}{A}\frac{d\ln(c_1 - c_2)}{dt} \tag{4-9}$$

Thus, the permeability of the coating can be determined from the slope of a linear relationship between logarithm $(c_1$-$c_2)$ and t.

Figure 4-19 shows the typical plots of $\ln(c_1$-$c_2)$ vs. time for different types of coating film for dissolved CO_2 permeation, where linear relationships are observed are all coatings. The coating permeability for dissolved CO_2 calculated from the slope of individual lines is shown in Table 4-4. There is the largest CO_2 permeability for European coating. Moreover, the permeability of PE is larger than that of HPCC and FBE. The comparison between MDPE and HDPE suggests that an increase in PE density decreases the coating permeability. There is the smallest CO_2 permeability for FBE.

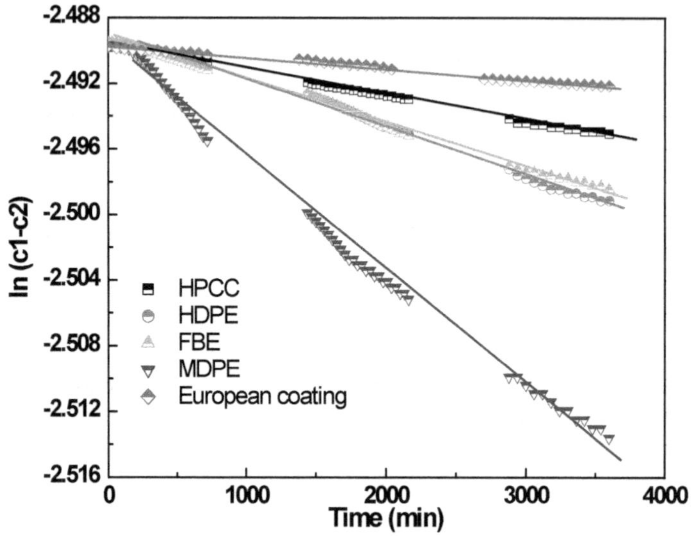

Figure 4-19. Plots of $\ln(c_1\text{-}c_2)$ vs. t for various coating membranes for dissolved CO_2 permeation measured at 73.4°F (23°C) [Cheng, et al., 2009].

Table 4-4. Permeability of Various Coatings to Dissolved CO_2 and O_2 at 73.4°F (23°C) [Cheng, et al., 2009].

Coating Type	CO_2 Permeability Coefficient (mol/atm m s)	O_2 Permeability Coefficient (mol/atm m s)
HPCC (0.92 mm)	7.9×10^{-12}	1.1×10^{-12}
FBE (0.12 mm)	1.7×10^{-12}	1.8×10^{-13}
MDPE (0.80 mm)	3.1×10^{-11}	2.4×10^{-12}
HDPE (0.80 mm)	1.1×10^{-11}	2.2×10^{-12}
European coating (5 mm)	4.1×10^{-11}	8.3×10^{-11}

The plots of $\ln(c_1\text{-}c_2)$ vs. time for the various coatings for O_2 permeation is shown in Figure 4-20. Similarly, the $\ln(c_1\text{-}c_2)$ decreases linearly with time after hundreds of minutes. The permeability of various coatings to O_2 calculated from the slope of the line is also included in Table 4-4. There is the smallest O_2 permeability in FBE, and the largest permeability for the European coating. The PEs have a larger O_2 permeability than HPCC. Furthermore, the increase in PE density would decrease its permeability to dissolved O_2.

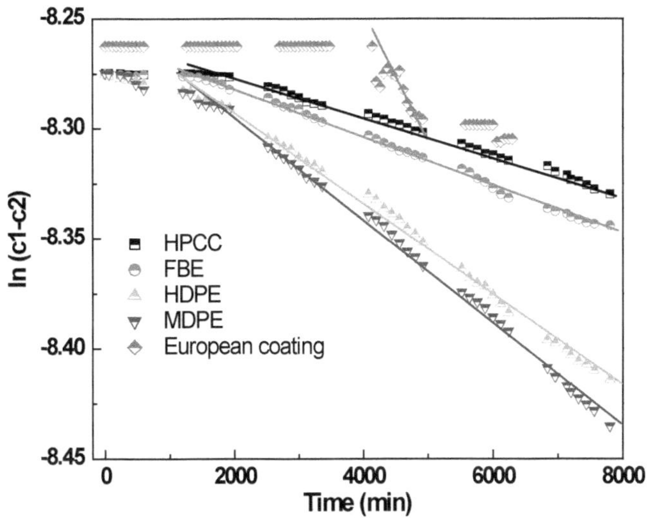

Figure 4-20. Plots of $\ln(c_1\text{-}c_2)$ vs. t for various coating membranes for dissolved O_2 permeation measured at 73.4°F (23°C) [Cheng, et al., 2009].

4.5. References

Baker, M. (2004) *Integrity Management Program–Stress Corrosion Cracking Studies*, Final Report, Office of Pipeline Safety TTO-8, Department of Transportation, Washington DC, USA.

Blanco, I., Cicala, G., Costa, M., Recca, A. (2006) Development of an epoxy system characterized by low water absorption and high thermomechanical performances, *J. Appl. Polym. Sci.* 100, 4880-4887.

Bollinger, J.C. (2015) The devil is in the details, *World Pipelines*, Special Issue on Coatings & Corrosion, 56-61.

Buchanan, R. (2013a) Cathodic shielding: absorbing the details, *World Pipelines* 13, 43-45.

Buchanan, R. (2013b) A critical review of industry codes and standards as they relate to electrically resistive coatings and what that means to shielding, *Corrosion'2013*, paper No. 2080, NACE, Houston, TX, USA.

Chen, X., Li, X.G., Du, C.W., Cheng, Y.F. (2009) Effect of cathodic protection on corrosion of pipeline steel under disbonded coating, *Corros. Sci.* 51, 2242–2245.

Cheng, Y.F. (2007) Fundamentals of hydrogen evolution reaction and its implications on near-neutral pH stress corrosion cracking of pipelines, *Electrochim. Acta* 52, 2661-2667.

Cheng, Y.F., Zhang, A., Fu, A. (2009) *Determination of permeability of multi-layer coatings to CO_2 and O_2*, Internal report, NOVA Chemicals, Calgary, AB, Canada.

Cheng, Y.F. (2013) Controversy contained: In-depth look at CP shielding, *World Pipelines* 13, 50-54.

Chike, K., Myrick, M., Lyon, R., Angel, S. (1993) Raman and near-infrared studies of an epoxy resin, *Appl. Spectrosc.* 47, 1631–1635.

DNV (2010) *Cathodic Protection Design, Recommended Practice*, DNV-RP-B401, Det Norske Veritas.

Ethridge, A., Yarmuch, M., Buchner, S. (2008) Permeation and corrosion testing of flexible steel pipe, *NACE International Northern Area Western Conference*, Edmonton, Canada.

Forsgren, A. (2006) *Corrosion Control through Organic Coatings*, Ch. 6 - Aging and Weathering of Paint, CRC Press, Boca Raton, FL, USA.

Fu, A.Q., Cheng, Y.F. (2010) Effects of alternating current on corrosion of a coated pipeline steel in a chloride-containing carbonate/bicarbonate solution, *Corros. Sci.* 52, 612-619.

Fu, A.Q., Cheng, Y.F. (2011) Characterization of the permeability of a high performance composite coating to cathodic protection and its implications on pipeline integrity, *Prog. Organ. Coat.* 72, 423-428.

Fu, A.Q., Cheng, Y.F. (2012) Effect of alternating current on corrosion and the effectiveness of cathodic protection of pipelines, *Can. Metall. Q.* 51, 81-90.

Fujita, H. (1968) *Diffusion in Polymers*, J. Crank and G.S. Park, Eds., Ch. 3, Academic Press, New York, USA.

Gabriel, L.H., Moran, E.T. (1998) *Service Life of Drainage Pipe*, Synthesis of Highway Practice 254, Transportation Research Board, Washington DC, USA.

George, G., Cole-Clarke, P., St. John, N., Friend, G. (1991) Real-time monitoring of the cure reaction of a TGDDM/DDS epoxy resin using fiber optic FTIR, *J. Appl. Polym. Sci.* 42, 643–657.

González, M.G., Baselga, J., Cabanelas, J.C. (2012) Applications of FTIR on epoxy resins-identification, monitoring the curing process, phase separation and water uptake, in: T. Theophanides (Ed.), *Infrared Spectroscopy Materials Science, Engineering and Technology*, InTech, USA.

Gulmine, J., Janissek, P., Heise, H., Akcelrud, L. (2002) Polyethylene characterization by FTIR, *Polym. Test.* 21, 557–563.

IPEX (2009) *PVC Chemical Resistance Guide*, 1st Edition, Mississauga, Ontario, Canada.

Jack, T., King, F., Cheng, Y.F., Worthingham, R. (2002) Permeable coatings and CP compatibility, *The 4th Inter. Pipeline Conf.*, Calgary, AB, Canada, p. 1889–1893.

Jackson, D. (2015) Coatings Q & A, *World Pipelines* 15, 46-49.

Jost, W. (1961) *Diffusion in Solids, Liquids and Gases*, Academic Press, New York, USA.

Krimm, S., Liang, C., Sutherland, G. (1956) Infrared spectra of high polymers. II. Polyethylene, *J. Chem. Phys.* 25, 549–562.

Kuang, D., Cheng, Y.F. (2014) Understand the AC induced pitting corrosion on pipelines in both high pH and neutral pH carbonate/bicarbonate solutions, *Corros. Sci.* 85, 304-310.

Kuang, D., Cheng, Y.F. (2015a) Study of cathodic protection shielding under coating disbondment on pipelines, *Corros. Sci.* 99, 249-257.

Kuang, D., Cheng, Y.F. (2015b) Probing potential and solution pH under disbonded coating on pipelines, *Mater. Perf.* 54, 40-45.

Kuang, D., Cheng, Y.F. (2015c) Effect of alternating current interference on coating disbondment and cathodic protection shielding on pipelines, NACE International Northern Area Western Conference, Calgary, AB, Canada, Feb. 23-26.

Lee, M.C., Peppas, N.A. (1993) Water transport in epoxy resins, *Prog. Polym. Sci.* 18, 947–961.

Leidheiser Jr., H., Granata, R.D. (1988) Ion transport through protective polymeric coatings exposed to an aqueous phase, *IBM J. Res. Dev.* 32, 582–590.

Masilela, Z., Pereira, J. (1998) Using the direct current voltage gradient technology as a quality control tool during construction of new pipeline, *Eng. Fail. Analy.* 5, 99-104.

Mayne, J.E.O. (1952) The mechanism of the inhibition of the corrosion of iron and steel by means of paint, *Off. Dig.* 24, 127-136.

Mertzel, E., Koenig, J.L. (1986) Application of FTIR and NMR to epoxy resins, in: *Epoxy Resins and Composites II*, Springer, Berlin, Heidelberg, Germany, 73–112.

Meure, S., Wu, D.Y., Furman, S.A. (2010) FTIR study of bonding between a thermoplastic healing agent and a mendable epoxy resin, *Vib. Spectrosc.* 52, 10–15.

Mijovic, J., Andjelic, S. (1995) A study of reaction kinetics by near-infrared spectroscopy. I. Comprehensive analysis of a model epoxy/amine system, *Macromolecules* 28, 2787–2796.

Mohan, S., Prabakaran, A. (1989) Infrared and laser Raman spectra of polyethylene and its normal coordinate analysis, *Asian J. Chem.* 1, 162-165.

Muncaster, N., Perrad, L. (2015) Cathodic disbondment or self-healing? *World Pipelines*, Special Issue on Coatings & Corrosion, 50-54.

Munger, C.G. (1994) Causes and Prevention of Paint Failure, in: *Good Painting Practice - Steel Structures Paint Manual*, Editors, J.D. Keane, D. Berger, H. Hower, B.R. Appleman, Volume 1, Chapter 23, Steel Structures Painting Council, Pittsburgh, PA, USA.

NACE (2002) *NACE Corrosion Engineers Reference Book*, Ed. 3, R. Baboian, Ed., NACE International, Houston, TX, USA.

Norsworthy, R. (2008) Coatings used in conjunction with cathodic protection-shielding vs. non-shielding pipeline coatings, *The 17th Inter. Corros. Cong.*, paper No. 4017, Las Vegas, NE, USA.

Norsworthy, R., "Is Your Pipeline Coating 'Fail Safe'?" *Pipeline and Gas Journal,* October 2006, pg. 62, Volume No. 233 Number 10.

Peabody, A.W. (2001) *Peabody's Control of Pipeline Corrosion*, 2nd Ed., NACE International, Houston, TX, USA, p. 6.

Perdomo, J.J., Song, I. (2000) Chemical and electrochemical conditions on steel under disbonded coatings: The effect of applied potential, solution resistivity, crevice thickness and holiday Size, *Corros. Sci.* 42, 1389-1415.

Pikas, J.L. (1996) Case histories of external microbiologically influenced corrosion underneath disbonded coatings, *Corrosion'1996*, paper No. 96198, NACE International, Houston, TX, USA.

Pratt, J. K., Mallozzi, M., D'Souza, A. (2011) Advances in Damage Resistant Coating Technology, *Corrosion'2011*, paper no. 11031, NACE International, Houston, TX, USA.

Revie, R.W., Uhlig, H.H. (2008) *Corrosion and Corrosion Control*, 4th Ed., John Wiley & Sons, Hoboken, NJ, USA.

Ritums, J.E., Mattozzi, A., Gedde, U.W., Hedenqvist, M.S., Bergman, G., Palmlof, M. (2006) Mechanical properties of high-density polyethylene and crosslinked high-density polyethylene in crude oil and its components, *J. Polymer Sci. Part B: Polymer Physics* 44, 641.

Roberge, P.R. (1999) *Handbook of Corrosion Engineering*, McGraw-Hill Professional, New York, NY, USA.

Sato, K. (1980) The internal stress of coating films, *Prog. Organ. Coat.* 8, 143-160.

Song, F.M., Kirk, D.W., Cormack, D.E., Wong, D. (2005) Barrier properties of two pipeline coating, *Corrosion'2005*, paper no. 05025, NACE, Houston, TX, USA.

Thomas, N.L. (1991) The barrier properties of paint coatings, *Prog. Org. Coat.* 19, 101–121.

US Code of Federal Regulations (2012a) Title 49 "Transportation", Part 192, Transportation of Natural and Other Gas by Pipeline, Minimum Federal Safety Standards, Washington DC, USA.

US Code of Federal Regulations (2012b) Title 49 "Transportation", Part 195, Transportation of Hazardous Liquids by Pipeline, Washington DC, USA.

Xu, L.Y., Cheng, Y.F. (2013) Effect of alternating current on cathodic protection on pipelines, *Corros. Sci.* 66, 263-268.

5 Coating Failure and Pipeline Stress Corrosion Cracking

5.1. Introduction

Stress corrosion cracking (SCC) describes service failure in engineering materials that occurs by slow, environmentally induced crack propagation. The observed cracking phenomenon is the result of combined and synergistic interactions of mechanical stress and corrosion reactions on susceptible materials, which are usually metals [Jones, 1992]. SCC has been attributed to a number of pipelines leaking and/or rupture events, resulting in energy loss, environmental damage, and sometimes, catastrophic consequences [National Energy Board, 1996]. Pipeline SCC occurs from a combination of factors: Environmental (coating, CP, temperature, soil), stress (hoop stress, longitudinal and/or residual stress, pressure fluctuations), and material (steel grade and chemistry, microstructure, metallurgical defects, surface roughness, welds) [Cheng, 2013].

In Canada, over 70% of pipeline SCC incidents were associated with PE-coating types, whereas 30% of incidents occurred with other coating types [National Energy Board, 1996]. No SCC failures have been reported on FBE-coated pipelines in over 40 years of experience [Been, 2011]. The occurrence of pipeline SCC and its mechanism are based on coating selections. Based on permeability to water, gases, and chemicals dissolved in the water, coatings can be categorized as permeable and impermeable (as previously discussed). The most common examples of permeable coatings are asphalt and FBE coatings. Impermeable coatings include PE-based coatings, such as PE tape or 3LPE.

Generally, pipeline SCC is categorized into two types (i.e., high-pH SCC and near-neutral pH SCC), based on the chemistry and pH of the electrolyte contacting the pipeline steel. The pH refers to the environment at the crack location, rather than the soil pH. Moreover, the pH and chemistry of the electrolyte generated under a disbonded coating depends on the type and properties of pipeline coatings and their failure modes as well as with the coating's compatibility with CP [Cheng, 2013].

5.2. Near-neutral pH SCC

Near-neutral pH SCC has been recorded as a degrading mechanism of pipelines until the 1980s. Three different pipelines operated by TransCanada Pipelines (Alberta, Canada) ruptured between 1985 and 1986, with the failures recognized as due to near-neutral pH SCC [National Energy Board, 1996]. The near-neutral pH SCC usually occurs beneath impermeable coatings that are disbonded from pipe steel. Groundwater penetrated and became trapped beneath the disbonded coating. The CO_2, which is generated from organic matter decay in soils, dissolves in the trapped electrolyte. Due to either the high resistivity of the ground water or the shielding effect of the coating (or both), CP current cannot reach the steel-pipe surface. A near-neutral pH solution environment could then be developed to support SCC.

5.2.1. Primary Features

Cracking environments associated with the near-neutral pH SCC of pipelines are anaerobic in nature. Diluted electrolyte trapped under disbonded coating is primarily made of bicarbonate ions, where dissolved CO_2 is mainly from decayed organic matters and geochemical reactions in the soil. At the same time, CP access is limited over a considerable period of time due to either shielding disbondment of the coating or a highly resistive soil. Therefore, pipeline steel is at its corrosion potential of about -760 mV and -790 mV(CSE) [National Energy Board, 1996]. Furthermore, although a distinct relationship is lacking between near-neutral pH SCC and temperature, cracking tends to occur in cold climates where CO_2 concentration in groundwater is high.

Both lab testing and field investigations found that cracks usually initiate and accumulate at sites on an external surface, where corrosion pits or surface flaws are present. Moreover, cracks associated with near-neutral pH SCC are wide and transgranular in nature [Cheng, 2013]. The cracks grow across grains in the steel, and the crack walls experience metal loss from corrosion. Generally, colonies of multiple parallel cracks are observed and are perpendicular to the direction of the highest stress on the external pipe surface. These cracks can vary in depth and length and grow in two directions (i.e., along the circumferential direction of the pipe and the axial direction on the pipe surface). About two-thirds of the incidents that involved near-neutral pH SCC have axially oriented cracking, which is also called axial SCC or A-SCC. The circumferential cracks generated in the near-neutral pH environment due to circumferential SCC (C-SCC) are usually in areas where secondary stresses are present due to reasons such as ground movement.

Factors including CP, coating, soil characteristics, microorganisms, stresses, and steel metallurgy contribute to the near-neutral pH SCC of pipelines. Particularly, the coating's property and its failure mode are critical to the generation of a near-neutral pH environment to support SCC. Statistically, the occurrence of near-neutral pH SCC on pipelines is mainly associated with PE-type coatings.

5.2.2. Coating Failure as a Contributing Factor

Coating failure has been recognized as one of the most important factors resulting in near-neutral pH SCC on pipelines. Ideally, when the coating has failed, the applied CP should protect steel from corrosion by penetrating the coating and reaching the steel pipe surface. However, if the coating (such as PE tape) is impermeable to the CP current, the CP becomes shielded from reaching the steel. As a result, the pipe steel finds its corrosion potential at disbonded areas, rather than CP potential in the trapped electrolyte (i.e., an anaerobic, diluted, bicarbonate solution).

Asphalt and coal-tar coatings may also disbond due to either poor surface preparations or weak bonding strength to the substrate. However, since these coatings tend to become saturated with moisture (or, if brittle, may break into pieces), the CP current can reach the pipe steel in the disbonded area. Thus, near-neutral pH SCC would not occur on pipelines beneath these coatings. Furthermore, the near-neutral pH SCC does not occur on FBE-coated pipelines because the FBE is permeable to CP current, which elevates the pH of the trapped electrolyte to an alkaline range by the CP-driven electrochemical reduction of dissolved oxygen or water.

The phenomenon of near-neutral pH SCC occurs mostly on pipelines coated with impermeable coatings, which are typically PE-tape coatings. Upon coating disbondment, water and chemicals may enter the disbonding crevice from gaps formed at tape overlaps. Carbon dioxide may also enter and get dissolved in water, generating a diluted, near-neutral pH bicarbonate solution with a pH range of 5.5 to 7.5. The pipe steel is at near-corrosion potential in the trapped electrolyte, giving rise to corrosion of the steel beneath the coating. During the corrosion process, corrosion products form and accumulate on the steel surface. Furthermore, stress corrosion cracks usually initiate in corrosion pits, which are generated from the preferential dissolution of metallurgical defects contained in pipe steel in the electrolyte [Liu et al., 2012a; Liu et al., 2012b]. Hoop stress resulting from operating pressure and pressure fluctuations provide the primary stress source for cracking processes. A surface catalytic effect due to mill scale and porous corrosion products on the pipe surface enhances hydrogen evolution, accompanying the steel dissolution. Thus, hydrogen atoms generated during corrosion may enter into the steel, contributing to the cracking process [Cheng and Niu, 2007]. The anaerobic condition of the near-neutral pH electrolyte beneath disbonded coating may favor the culture and growth of microbiological populations, such as sulfate-reducing bacteria (SRB) [National Energy Board, 1996]. SRB can be a catalyst for hydrogen atoms permeating steel through generated sulfide ions via the electrochemical reduction of sulfate ions [Little et al., 2006] and the poisoning effect on hydrogen permeation. Thus, the involvement of hydrogen in near-neutral pH SCC processes is enhanced in the presence of SRB.

In summary, the critical factor for facilitating near-neutral pH environments to support SCC on pipelines is the shielding effect of the disbonded coating on CP current access to the substrate pipe steel.

5.2.3. Electrochemical Aspects of Pipeline SCC in Thin Layers of Near-neutral pH Electrolyte beneath Disbonded Coating

While extensive research has been conducted to study the electrochemical aspects of pipeline SCC in simulated electrolytes trapped beneath disbonded coating [Cheng, 2013], all of them were performed by immersing steel specimens in bulk solutions. In reality, electrolyte trapped between the coating and the pipe steel is very thin, usually at tens or hundreds of microns in thickness, especially at the early stage of coating disbondment. Electrochemical corrosion of steels in a thin layer of electrolyte is distinctly different from that in bulk solutions. For example, a small ohmic potential drop and non-uniform current distributions in the thin electrolyte layer are expected to significantly affect the corrosion reaction of the steel [Nishikata et al., 1995]. Currently, there has been limited work with a thin, aqueous layer to reproduce the actual condition encountered beneath disbonded pipeline coating. The lack of investigations under the actual corrosive environment is primarily due to the experimental difficulty with measuring corrosion and SCC in the electrolyte's thin layer.

The scanning Kelvin probe (SKP) technique provides a promising method that enables measurements of the electrochemical corrosion behavior of metals in a thin layer of electrolyte [Stratmann and Streckel, 1990a; Stratmann and Streckel, 1990b; Stratmann et al., 1990]. Moreover, the SKP can be used in coating studies with the provision of unique results. SKP's fundamentals and application in coating characterizations are described in Chapter Seven. Figure 5-1 shows a schematic diagram of the apparatus for polarization curve measurements using an SKP in a thin aqueous solution layer trapped at the steel/coating interface [Fu et al., 2009]. An X70-steel working electrode is covered with a solution layer, with a thickness measured by a three-dimensional imaging device. An FBE film with a 100-μm thickness is placed above the solution. The Kelvin probe with a 500-μm tungsten (W) tip serves as the reference electrode and is set directly above the coating. The distance between the probe tip and the coating surface is controlled at 250 μm. A Pt wire is used as the counter electrode. During testing, a current is applied with a potentiostat on the targeted coated-steel electrode, and the resulting Volta potential difference between the steel and the Kelvin probe is measured through a M370 scanning electrochemical workstation. The polarization curve is then measured starting from corrosion potential, cathodically polarized first, then the anodic portion by altering the current.

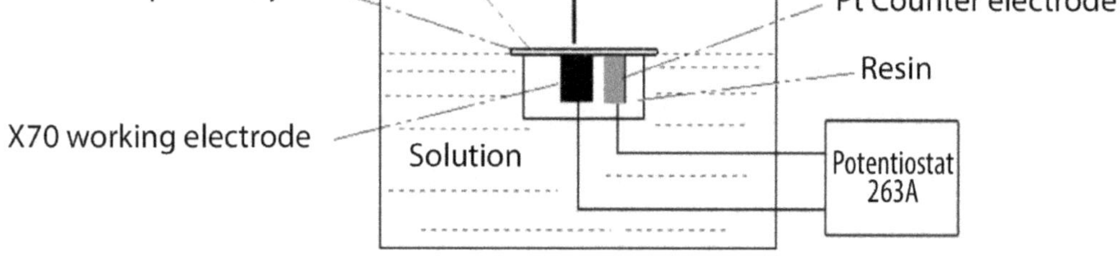

Figure 5-1. Schematic diagram of the apparatus for polarization curve measurements on coated steel using an SKP in a thin aqueous solution layer trapped at steel/coating interface [Fu et al., 2009].

Figure 5-2 shows polarization curves measured on X70 steel after immersion for 1 h and 24 h in a layer of near-neutral pH bicarbonate solution with various thicknesses. The effect of the solution-layer thickness on the anodic polarization curve is apparent. When the solution layer is 60 μm thick, the steel can be passivated with a stable passive potential range from approximately 0 to 0.5 V(SHE). This feature is not observed in the polarization curves measured at 90-μm and 140-μm solution layers. When the solution-layer thickness is increased to 90 μm, passivity of the steel becomes less stable after 1 h of immersion compared to that measured under a 60-μm solution layer. After 24 h of immersion, passivity could not be maintained. In the 140-μm solution layer, passivity is not observed after both 1 h and 24 h of immersion, and the steel shows an active dissolution state. Furthermore, the anodic current density increases when the immersion time increases from 1 h to 24 h for all curves. Compared with the polarization curve measured on the same steel in a bulk near-neutral pH bicarbonate solution (as shown in Figure 5-3), there is a similar feature to that measured in the 140-μm solution layer (i.e., there is no passivity observed), and the steel is in an active dissolution state.

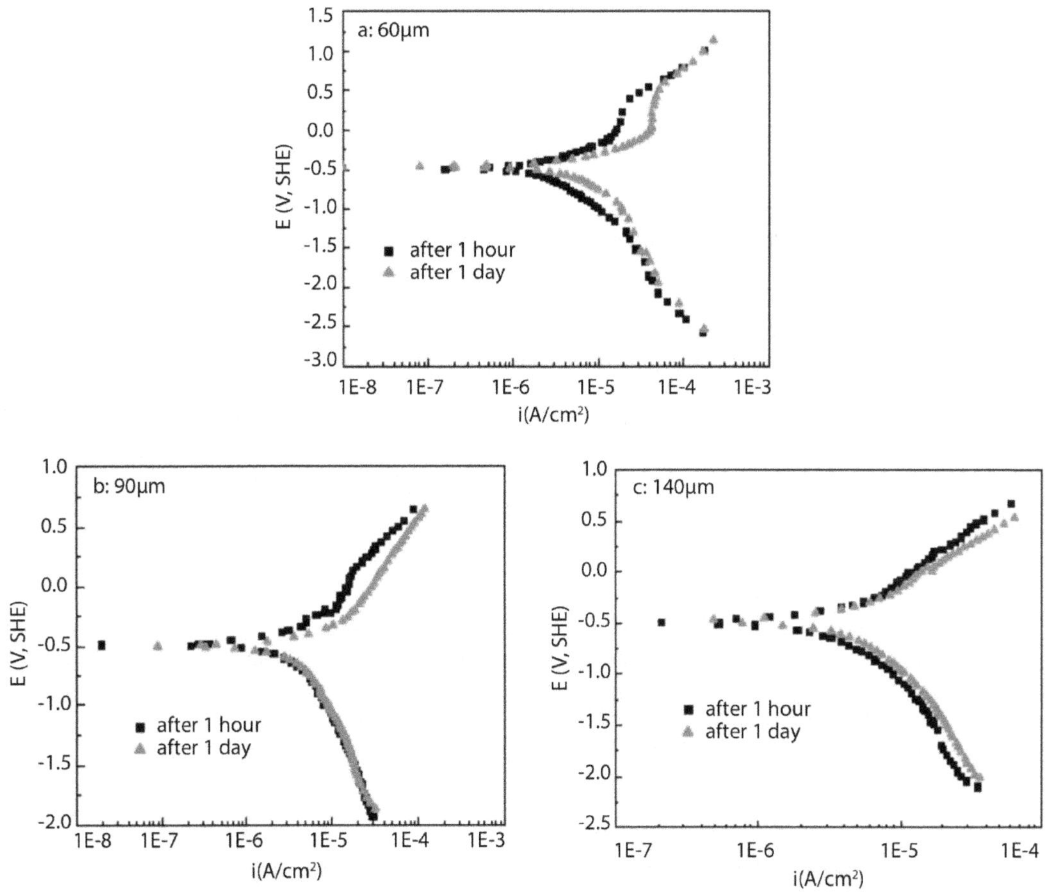

Figure 5-2. Polarization curves measured on X70 steel after 1 h and 24 h of immersion in a thin layer of near-neutral pH bicarbonate solution with various thicknesses, respectively [Fu et al., 2009].

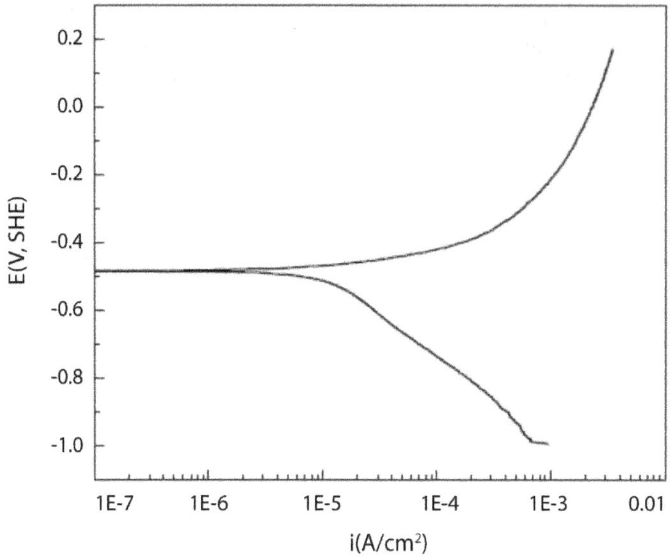

Figure 5-3. Polarization curve of X70 steel measured in a bulk near-neutral pH bicarbonate solution under the same condition as that in Figure 5-2 [Fu et al., 2009].

It has been demonstrated [King et al., 2000; Cheng and Niu, 2007] that the anodic and cathodic reactions of pipeline steels in a deoxygenated, near-neutral pH bicarbonate solution are iron oxidation and water reduction, respectively. A layer of loose, porous corrosion products (primarily $Fe(OH)_2$) could form on the steel surface. Due to the solution's diluted nature and the low testing temperature, $FeCO_3$ scale cannot be formed, although 5% CO_2 is purged continuously into the solution to maintain a near-neutral pH of about 6.5. It has been generally accepted that the electrochemical corrosion of pipeline steels in the near-neutral pH bicarbonate solution is in an active dissolution state [Cheng, 2010; Cheng, 2013], as shown in Figure 5-3.

However, measurement results show that, under a thin layer of a near-neutral pH bicarbonate solution (such as that of 60 μm in thickness), passivity can be established for the steel. It is attributed to the concentration of Fe^{2+} ions generated by corrosion of the steel, reaching a saturation state in the thin solution layer. Although the amount of CO_3^{2-} generated in the system is very low (diluted solution), it is still likely to exceed the solubility of $FeCO_3$ so that the iron carbonate scale can be precipitated once the Fe^{2+} concentration achieves a saturation or even super-saturation by:

$$Fe^{2+} + CO_3^{2-} \rightarrow FeCO_3 \qquad (5\text{-}1)$$

With an increased solution-layer thickness, it becomes more difficult for the concentration of Fe^{2+} to reach saturation. It is apparent from Figure 5-2 that, when the solution layer increases to 140 μm, the steel cannot be passivated any more, and the measured polarization curve is similar to that measured in the bulk solution (Figure 5-3).

Because extensive studies have shown that pipeline steels are in an active dissolution state in near-neutral pH environments, it is believed that the dissolution-based mechanism associated with the passive film rupture and repair does not apply to near-neutral pH SCC of pipelines. For example, it was proposed [Parkins, 2000] that hydrogen is involved in near-neutral pH SCC of pipelines and accompanies anodic dissolution at the crack tip. Previous work with the electrochemical behavior of pipeline steels in near-neutral pH solutions has been conducted in bulk aqueous solutions. The present work, in its first edition, demonstrates that, when a thin layer of near-neutral pH electrolyte forms on the steel surface, a stable passivity develops on the steel. In reality, the trapped electrolyte beneath a disbonded coating is usually very thin, especially at the disbondment's early stage. Moreover, the space beneath the coating disbondment is often very small, limiting the free exchange between trapped electrolytes and the outside soil environment. It is therefore reasonable to assume that pipeline steel is actually covered with a passive film in a thin layer of electrolyte over a probable long time period, during which the steel is in a passive state, rather than an active dissolution state. Therefore, the conventional acknowledgement that pipeline steels are in an active dissolution state under near-neutral pH environment is questionable.

This finding is of great importance for the machinistic understanding of pipeline SCC in near-neutral pH environments. It advances our knowledge of this important phenomenon, overthrowing some of the accepted methods and requiring new technologies to manage it. For example, once near-neutral pH stress corrosion cracks initiate on pipelines (where the steel is under a thin layer of electrolyte and is thus in passivity), the role of hydrogen becomes less important due to the inhibition of the surface passive film on hydrogen that permeates steels [Song et al., 1990]. The well-accepted proposition about hydrogen involvement in pipeline SCC seems unreliable. As a consequence, models based on the synergism of hydrogen and anodic dissolution at the crack tip to predict the SCC propagation rate and remaining life of pipelines in near-neutral pH environments [Parkins et al., 1994; Baker, 2004; Cheng, 2007] should be modified or even abandoned.

5.3. High-pH SCC

High-pH SCC of pipelines is a classical SCC, originally noted in gas-transmission pipelines. The first documented case of SCC resulting in a gas release, explosion, and fire with several fatalities in Louisiana, U.S. in the mid 1960s was caused by high-pH SCC [National Energy Board, 1996]. To date, high-pH SCC has been recognized on pipelines in the United States, Australia, Iran Iraq, Italy, Pakistan, Saudi Arabia, and in many other countries. It has become a worldwide problem affecting pipeline integrity and safe operation.

5.3.1. Primary Features

High-pH SCC happens externally on pipelines where the electrolyte, primarily concentrated carbonate-bicarbonate solution, in contact with the pipe steel under disbonded coating has a pH of 9-11. Generation of this high pH electrolyte results from CP current permeating the coating, driving cathodic reduction of water or dissolved oxygen to generate hydroxyl ions, which interact with carbon dioxide generated from decaying organic matters in soils. The high-pH electrolyte causing pipeline SCC does not refer to nearby soil chemistry.

High-pH SCC is sensitive to temperature and occurs more frequently at higher temperature locations almost above 100°F (38°C). This is why there is a higher probability of SCC instantly downstream from the compressor stations where operating temperatures may reach 149°F (65°C) [National Energy Board, 1996]. Thus, the number of failures decreases with increased distance from compressor/pump stations and with lower pipe temperatures.

The fracture mode of high-pH SCC tends to be intergranular, often with small branches. The cracks are usually narrow and tight, with almost no evidence of corrosion on the crack wall. A thin oxide layer is formed around the crack, providing protection in the concentrated carbonate-bicarbonate environment. However, due to changes in loading or cyclic loading from internal pressure or pressure fluctuations, there is a stress/strain concentration at the crack tip, resulting in breakage of the oxide film. This causes crack extension due to corrosion. Moreover, there exists extensive secondary cracks between grains as a result of this failure's branched nature. The high-pH stress corrosion cracks are primarily axial cracks, with very few transverse cracking cases [National Energy Board, 1996].

5.3.2. Coating Failure as a Contributing Factor

High-pH SCC requires an alkaline carbonate/bicarbonate solution with a pH range of 9–11, which results from the synergism between the CP current and the chemistry of the trapped electrolyte. The coating properties are critical for generating the water chemistry (i.e., the coating should allow the CP current to penetrate through it to cause the electrochemical reduction of water or dissolved oxygen to generate an alkaline environment). Thus, occurrence of the high-pH SCC of pipelines is always associated with a permeable coating (such as asphalt, coal tar, etc.) that is disbonded from the pipe steel.

Another requisite for initiating high-pH SCC is that the pipe steel potential must lie strictly within a well-defined range. The range of cracking potentials is temperature-dependent and is typically from -0.60 to -0.75 V(CSE) at room temperature. Therefore, the penetration of CP current through the coating is required to generate the high-pH environment, and the CP potential is then lost so that the steel potential falls into the cracking range. Factors such as the deposit of insolvable salts (e.g., cadmium carbonate, magnesium carbonate, etc.) on the coating can shield partially the CP current, causing the potential of the steel to drop into the cracking potential range.

The resistance of a coating to disbondment and the type of surface preparation used with the coating are two other factors affecting the resistance of pipelines to high-pH SCC [Beavers, 1992; Beavers et al., 1993a; Beavers et al., 1993b]. A coating's ability to resist disbondment is a primary property and affects almost all forms of external pipeline corrosion, including SCC. Coatings with good adhesion are generally resistant to mechanical actions of the soil from wet/dry cycles and freeze/thaw cycles. They can also resist water transmission and cathodic disbondment.

5.3.3. Electrochemical Aspects of Pipeline SCC in Thin Layers of High pH Electrolyte beneath Disbonded Coating

Potentiodynamic polarization curves were measured on X70 steel in a thin layer of a high-pH carbonate-bicarbonate electrolyte using an artificial experimental setup as shown in Figure 5-4 (where an FBE coating film with a thickness of 50 μm was applied to the steel surface using an instant adhesive). The gap between the coating and the steel, with thicknesses of 30 μm, 150 μm, and 250 μm, respectively, was prepared to simulate coating disbondment. A three-electrode cell was placed in a closed chamber to minimize an evaporation-induced change of the solution layer thickness under the disbonded coating. Test solutions contained carbonate/bicarbonate solutions with various concentrations, including 0.05 M Na_2CO_3 + 0.1 M $NaHCO_3$ (pH 9.6), 0.25 M Na_2CO_3 + 0.5 M $NaHCO_3$ (pH 9.4), and 0.5 M Na_2CO_3 + 1 M $NaHCO_3$ (pH 9.4). For convenience, the solutions were named as low, intermediate, and high concentrations of carbonate/ bicarbonate solution, respectively.

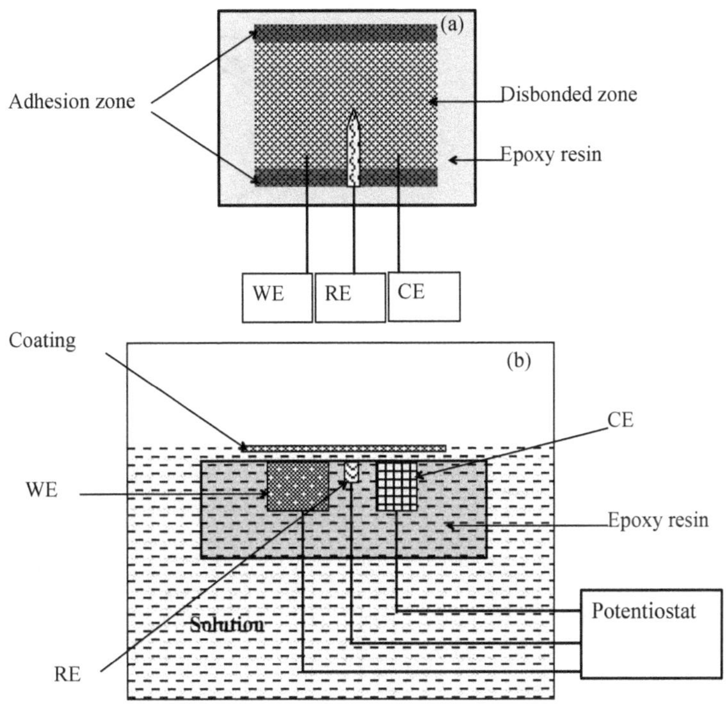

Figure 5-4. Schematic diagram of the coated steel specimen and electrode arrangement: (a) top view; (b) side view, where WE, CE, and RE are working electrode (X70 steel), counter electrode (a Pt plate), and reference electrode (SCE) [Fu and Cheng, 2010].

Effect of the solution layer thickness. Figure 5-5 shows the polarization curves measured on X70 steel in the low, intermediate, and high concentrations of carbonate/bicarbonate solution with various thicknesses. There is a significant influence of the solution-layer thickness on the electrochemical polarization behavior of the steel. In the low-concentration solution (as shown in Figure 5-5a), the steel can be passivated with a high active-passive transition current density. With the decrease of the solution-layer thickness, the transition current density decreases. When the solution layer reduces to 30 μm in thickness, the active-passive transition phenomenon disappears with the stable passivity developed on the steel. Although there is no well-defined dependence of the passive current density (i_p) and pitting potential (E_{pit}) on the solution layer thickness, there is the smallest i_p and the most negative E_{pit} measured in the solution with a thickness of 30 μm. In the intermediate concentration solution (Figure 5-5b), the passive current density is independent of the solution-layer thickness. The E_{pit} shifts negatively when the solution-layer thickness decreases from a bulk solution to 30 μm. With a further increase of the solution concentration, passivity developed on the steel in all solutions, as seen in Figure 5-5c.

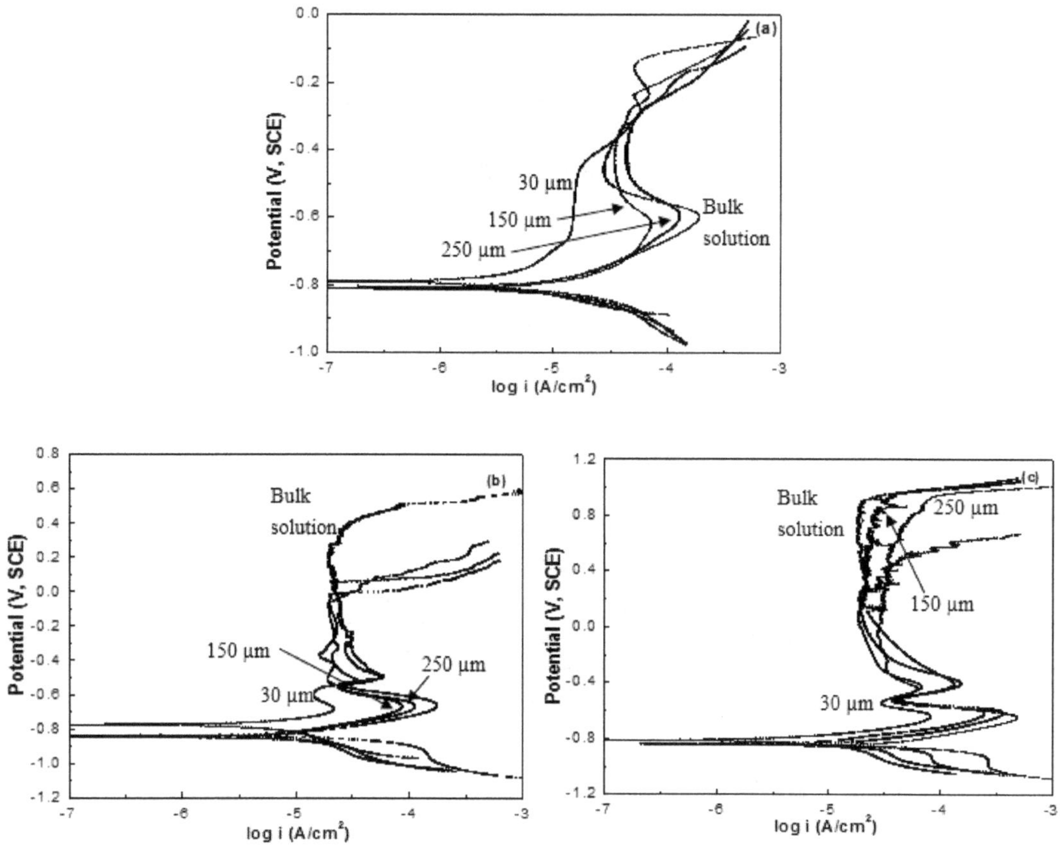

Figure 5-5. Polarization curves of X70 steel in (a) low, (b) intermediate, and (c) high concentrations of carbonate/bicarbonate solution with various thicknesses, respectively [Fu and Cheng, 2010].

A comparison of the polarization curves measured in Figure 5-5 shows that a decreasing solution-layer thickness would enhance passivity of the steel in a low concentration of the carbonate-bicarbonate solution. However, the E_{pit} shifts negatively, which means that the pitting susceptibility of the passivated steel increases. With an increase of the solution concentration, steel passivity is enhanced, indicated by the positively increasing E_{pit} and a wide passive potential range. However, the role of the solution layer thickness in passive current density becomes unapparent.

Affect of applied CP. In addition to the thickness of the trapped solution layer under disbonded coating, the applied CP would also affect the electrochemical polarization behavior of the steel in the solution, as the CP current permeates the coating. Figure 5-6 shows the polarization curves of X70 steel measured in low, intermediate, and high concentrations of a carbonate/bicarbonate solution with a 30-μm thickness as a function of CP duration, where the applied CP potential is -1.0 V(SCE). A pre-applied CP causes a negative shift of corrosion potential and an increase of the anodic current density. However, the duration of CP does not change the polarization feature of the steel.

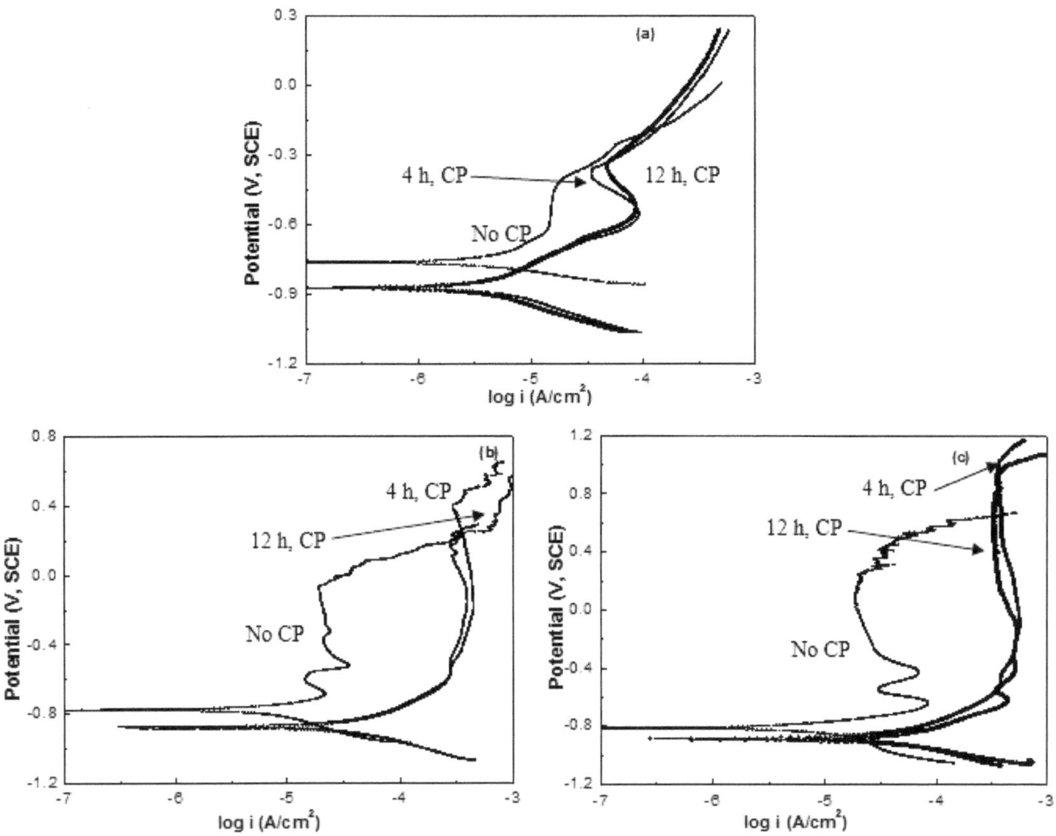

Figure 5-6. Polarization curves of X70 steel in (a) low, (b) intermediate, and (c) high concentrations of carbonate/bicarbonate solution with a 30 μm thickness as a function of the CP duration [Cheng and Fu, 2010].

The cathodic reaction occurring under applied CP can be determined by electrochemical thermodynamic calculations. The potential for hydrogen evolution (E_H) is:

$$E_H(SCE) = -0.0592 \times pH - 0.241 \tag{5-2}$$

The average pH of the solutions in this work is 9.5. The potential for hydrogen evolution is about -0.81 V(SCE). At a CP of -1.0 V(SCE), the hydrogen evolution dominates the cathodic process:

$$H_2O + e \rightarrow H + OH^- \tag{5-3}$$

The generated hydrogen atoms may penetrate the steel. The measurements above show that, upon application of CP, the corrosion potential of the steel shifts negatively and the passive current density increases, indicating a hydrogen-enhanced activity of the steel [Li and Cheng, 2007]. Similarly, upon application of CP, the solution pH would change due to the generation of OH^- by electrochemical cathodic reactions. At a CP potential (E_{CP}) that can cause hydrogen evolution, the solution pH is dependent on E_{CP} by:

$$pH = -(E_{CP} + 0.241)/0.0592 \tag{5-4}$$

As the CP potential shifts negatively, the pH of the solution increases, and the concentration of hydrogen ions decreases. The corrosion potential of the steel in the absence and presence of CP (E_{corr} and E_{corr}^{CP}) can be expressed as:

$$E_{corr} = E^0_{Fe^{2+}/Fe} + \frac{2.303RT}{2F}\log\frac{[Fe^{2+}][H_0^+]^{1/2}}{[H_{ad}^0]^{1/2}} - E_{ref} \tag{5-5}$$

$$E_{corr}^{CP} = E^0_{Fe^{2+}/Fe} + \frac{2.303RT}{2F}\log\frac{[Fe^{2+}][H_{CP}^+]^{1/2}}{[H_{ad}^0]^{1/2}} - E_{ref} \tag{5-6}$$

where $E_{Fe2+/Fe}^0$ is the standard equilibrium potential of iron, R is ideal gas constant, T is temperature, F is Faraday's constant, $[Fe^{2+}]$ is the concentration of ferrous ions, $[H_0^+]$ and $[H_{CP}^+]$ are the concentrations of hydrogen ions in the solution in the absence and presence of CP, respectively, $[H_{ad}^0]$ is the sub-surface concentration of adsorbed hydrogen atoms in the charged steel, and E_{ref} is the potential of reference electrode. Since $[H_0^+] > [H_{CP}^+]$, the corrosion potential of the steel under CP is more negative than that without CP.

Moreover, when hydrogen-charged steel is subject to anodic polarization, the anodic current density is greater than that measured on uncharged steel. The oxidization of the adsorbed hydrogen atoms contributes to the increasing anodic current density. Furthermore, the produced H^+ decreases the solution pH under the disbonded coating, which would also increase the dissolution current density of the steel. The similar results were observed in a previously published work [Yu et al., 2001].

Effect of stress. Buried pipelines are pressurized infrastructures. The hoop stress generated from the operating pressure always exists on buried pipe steel. A previous work [Zhang and Cheng, 2010; Tang and Cheng, 2011] established the correlation between the steel corrosion rate and various stress levels in near-neutral pH and high pH solutions, which, without exception, refer to bulk solutions. The knowledge about the effect of applied stress on corrosion and localized corrosion in thin layers of electrolytes has been limited. Figure 5-7 shows the polarization curves of X70 steel measured in low, intermediate, and high concentrations of the carbonate/bicarbonate solution with a 30-μm thickness as a function of applied stress. A similar effect of applied stress exists on polarization curves measured in all solutions with various stress levels (i.e., the passive current density increases and the passive potential range decreases with the applied stress).

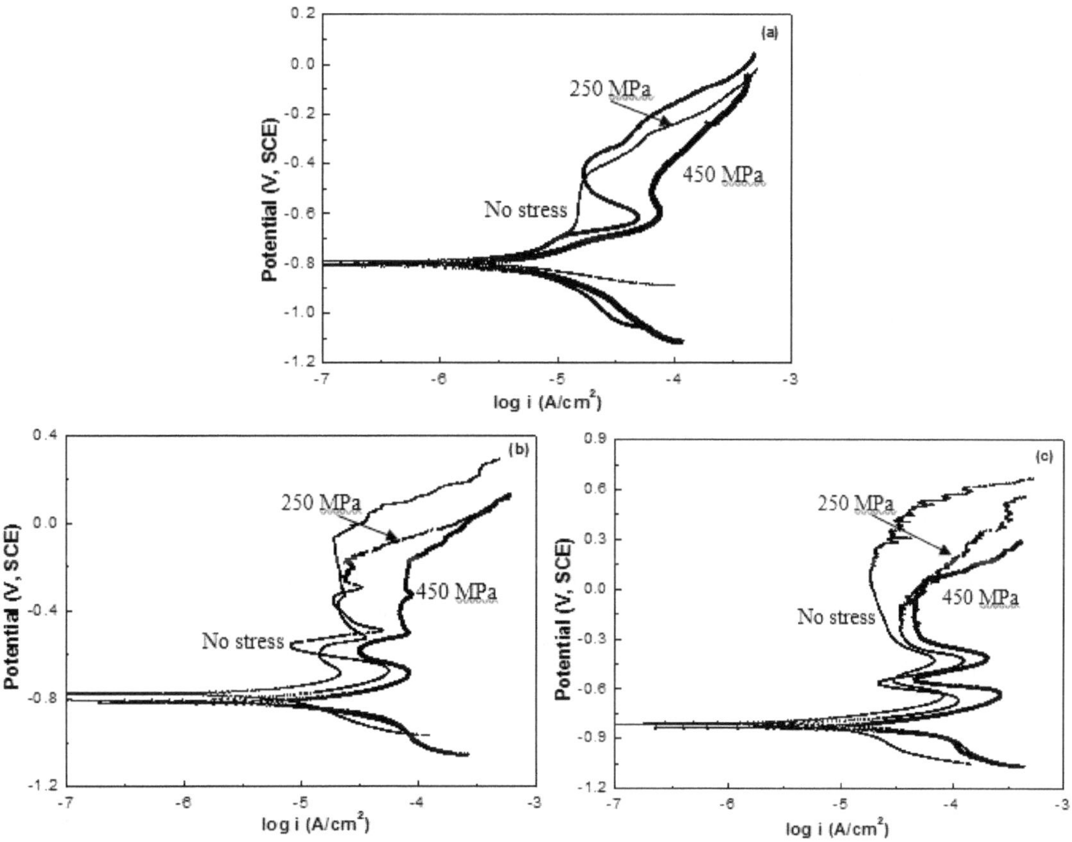

Figure 5-7. Polarization curves of X70 steel measured in (a) low, (b) intermediate, and (c) high concentrations of carbonate/bicarbonate solution with a 30 μm thickness as a function of applied stress [Fu and Cheng, 2010].

It has been acknowledged [Xu and Cheng, 2013] that applied stress, especially plastic stress, enhances anodic dissolution of steels. The passive film formed on the steel surface is prone to rupture when subjected to a sufficiently high stress, resulting in an increase of the anodic current density. The E_{pit} is highly related to the properties of the passive film. Generally, a passive film formed on steels and under a tensile stress is more active or weaker than that formed in the absence of stress. Thus, the E_{pit} usually shifts negatively when a stress is applied, resulting in a small passive potential range.

5.3.4. Modeling of the Occurrence of High-pH SCC on Pipelines

Occurrence of high-pH SCC on pipelines is usually associated with a concentrated, high-pH carbonate/bicarbonate solution trapped beneath disbonded coatings, which are permeable to CP current. Various factors, including coating degradation modes, CP, stress, etc., contribute to the development of a solution chemistry and electrochemistry that support pipeline SCC. The processes of pipeline SCC in high-pH environments can be illustrated by the following stages.

Development of a high-pH carbonate/bicarbonate electrolyte under disbonded coating. Pipeline integrity is maintained by both a protective coating and CP. For permeable coatings such as FBE, the CP current penetrates the coating and reaches the steel surface. When water is trapped beneath the disbonded coating, electrochemical reduction reactions are driven by CP to generate OH^- and elevate the pH of the trapped electrolyte. Simultaneously, CO_2 generated by decay of various organic matters in soil permeates and then dissolves in the electrolyte. At a pH of about 9-11, a concentrated carbonate/bicarbonate electrolyte is formed due to the hydrolysis of H_2CO_3 and HCO_3^- in the alkaline environment [King et al., 2000].

Initiation of corrosion pits in thin layers of electrolyte beneath coating. When the coating is initially disbonded from the pipe steel, the crevice generated beneath the disbonded coating is narrow, and the trapped solution experiences an elevation of the carbonate/bicarbonate concentration due to the permeation and dissolution of CO_2. At this stage, in both low and high concentrations of the solution, there is a lower E_{pit} in a thin solution layer than in a thick solution layer. In the presence of Cl^-, corrosion pits, which are usually the incubators to stress-corrosion cracks, initiate early in coating disbondment, where the trapped solution layer is usually thin and the solution concentration is relatively low. With the increase of the solution-layer thickness and the solution concentration, the E_{pit} of the steel shifts positively, and the resistance of the steel to pitting corrosion increases. Furthermore, the applied CP on the coated steel increases the passive current density, making the formed passive film more active. As a result, it would be easy for the steel to be attacked by chloride ions, resulting in the initiation of pitting corrosion.

Transition of corrosion pits towards cracks. It is believed [Van Boven et al., 2007] that corrosion pits act as stress raisers for SCC initiation. Stresses exerted on pipelines can facilitate the transition of pits towards cracks. An application of tensile stress increases the passive current density in the solution, resulting in the steel becoming more susceptible to pitting corrosion. It is reasonable to assume that stress exerted on pipelines from internal pressure and pressure fluctuations enhances the transition of corrosion pits towards cracks.

Propagation of stress-corrosion cracks. From the perspective of cracking kinetics, crack propagation requires steel to be in a certain potential range, which neither allows formation of a permanently stable passive film nor supports a constant active corrosion dissolution at the crack tip. The propagation rate of cracks depends on the competition between the rate of film growth and that of film rupture. The crack tip works as a stress raiser when the tensile stress is applied, while the crack wall is free of appreciable stress concentration. As there is a distinct difference of anodic current densities of the steel in the presence and absence of applied stress, the crack tip (compared with crack walls) is expected to be subject to a large dissolution rate due to periodic film rupture resulting from the applied stress. As a consequence, the dissolution of the crack tip will be further enhanced.

When pipelines are subjected to seasonal wet/dry cycles, the CP current may reach the steel surface in wet seasons, creating a high-pH carbonate/bicarbonate solution under disbonded regions [Perdomo et al., 2001]. In dry seasons, water evaporation generates concentrated carbonate/ bicarbonate solutions, and CP becomes inaccessible to the steel due to the increased soil resistivity. As a result, the potential of pipeline steel will be shifted in a positive direction and may locate in the potential range in which SCC is susceptible. Research results shown above indicate that the stability of the passive film formed on the steel in the absence of CP is better than that formed under CP. Therefore, crack growth is quite possible to occur in wet seasons rather than a dry season.

5.4. Modeling Solution Chemistry Developed beneath Disbonded Coating to Support Pipeline SCC

5.4.1. High-pH Solution Chemistry

The high-pH solution chemistry usually develops beneath disbonded permeable coatings, such as coal tar, asphalt, or FBE, which are compatible with CP. Once the coating is disbonded and when water permeates it, electrochemical reactions driven by CP occur on the pipe steel to generate a solution environment that is different from the surrounding soil. The electrochemical cathodic reduction of dissolved oxygen or water at CP potentials leads to the generation of hydroxyl ions, which raises the local pH (up to 9-11) [Cheng, 2013]. The CO_2 created by the decay of organic matters in the soil is dissolved in the solution and generates concentrated carbonate-bicarbonate electrolytes. The potential of the pipe steel is cathodically polarized to more negative values than its corrosion potential.

In addition to the electrochemical reaction, the mass transfer of chemical species and gases such as CO_2 and O_2 through the coating affects the solution chemistry under the permeable coating. Thus, the ground water composition also plays an important role in the generation of the solution environment. Ground water with high concentrations of Ca^{2+} and Mg^{2+} is more prone to the precipitation of insoluble carbonate salts, which may decrease the coating's permeability.

King model. King et al [King et al., 2004] developed a mathematical model to predict the generation and evolution of the environment under a disbonded permeable coating with a complete consideration of the synergistic effects of CP, coating, and soil environment. The model couples electrochemical reactions on the surface of the pipe with the transport of ions to and from the pipe surface through the permeable coating and the surrounding soil. Figure 5-8 shows the schematic diagram of the processes and environment included in the model.

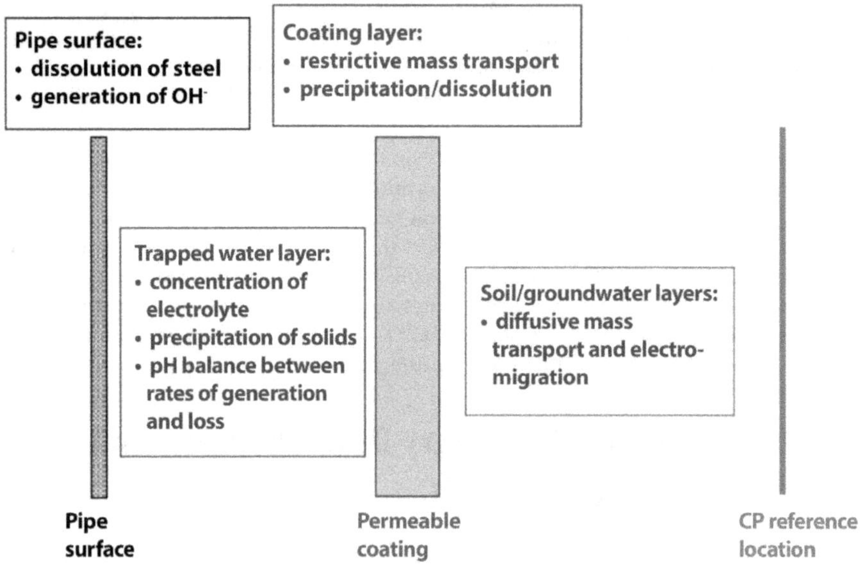

Figure 5-8. Schematic diagram of the processes and environment included in the permeable coating model [King et al., 2004].

The model was based on a conceptual understanding of the chemical and electrochemical processes occurring on or near the pipe surface under CP-compatible coatings. Figure 5-9 shows the chemical and electrochemical reactions under consideration. A total of 17 species are included in the model (plus water), consisting of 14 dissolved species (i.e., dissolved CO_2(aq), carbonic acid H_2CO_3, bicarbonate HCO_3^-, carbonate CO_3^{2-}, protons H^+, hydroxide OH^-, ferrous ions Fe^{2+}, ferrous carbonate complex ions $Fe(CO_3)_2^{2-}$, sodium ions Na^+, calcium ions Ca^{2+}, magnesium ions Mg^{2+}, chloride ions Cl^-, and sulfate ions SO_4^{2-}) as well as three precipitated species (i.e., iron carbonate $FeCO_3$, calcium carbonate $CaCO_3$, and magnesium carbonate $MgCO_3$). The concentrations of each species (except for H_2O) vary in time and space due to mass transport processes and chemical and electrochemical reactions.

The input data in the model include site-specific parameters to define the characteristics of the three layers (i.e., trapped water, coating, and soil layers) in Figure 5-8. Each layer is characterized by its thickness, the initial condition such as porosity, and the tortuosity factor. The trapped water layer thickness is equal to the disbondment thickness (i.e., the gap between the pipe surface and the underside of the disbonded coating). For disbonded asphalt or coal-tar enamel coatings, the thickness of the trapped water layer might be of the order of 0.1–1 mm. For an FBE coating, a trapped water-layer thickness of 0.01–0.1 mm could be used [King et al., 2004]. The coating layer thickness is chosen to represent the type of coating being simulated, and the soil layer thickness represents the distance from the outside of the coating to the location at which CP data are measured. Other parameters that characterize soil type include the concentrations of the different dissolved species as listed (other than Fe^{2+} and $Fe(CO_3)_2^{2-}$). The CP system is characterized by inputting either the cathodic potential or current density.

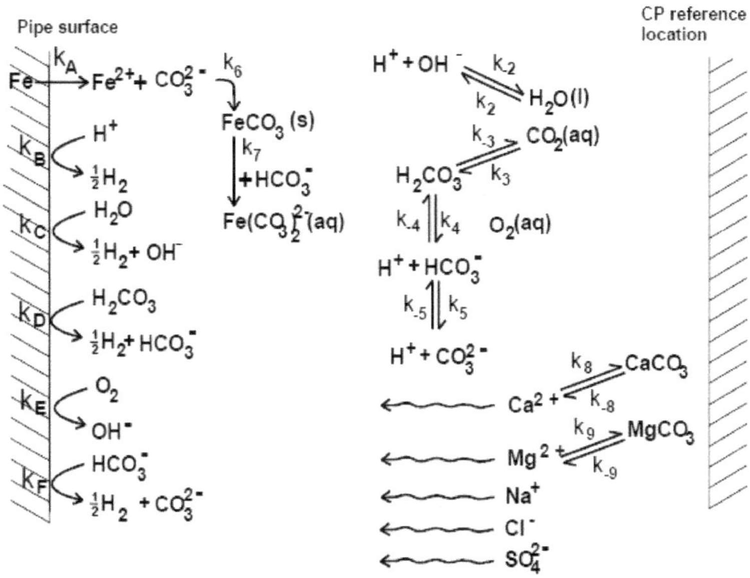

Figure 5-9. Chemical and electrochemical reactions considered in the permeable coating model [King et al., 2004].

The output data include spatial and temporal variations of the concentrations of each species and the rates of the electrochemical and chemical reactions listed in Figure 5-9. Specifically, the output data relevant to the corrosion and SCC behaviour of the pipe include the time dependences of the pipe steel potential, the corrosion rate, the thickness and porosity of the precipitated $FeCO_3$ film on the pipe surface, and solution pH. The output data also include those relevant to the coating conditions, trapped water chemistry, and CP performance.

A 30-day simulation period by the model shows that the solution chemistry developed under the disbonded permeable coating does not support high-pH SCC on pipelines. For example, the high-pH SCC is associated with concentrated carbonate/bicarbonate solutions (typically between 0.1 mol/L and 1.0 mol/L) with a pH in the range between 9.3 and 10.5 and potentials (at room temperature) of -0.60 to -0.75 V (CSE). However, according to the model, the trapped-water pH approaches a steady-state value of pH 12.2 after 1 month. The increase in pH under the coating is accompanied by an increase in the Na^+ concentration. Thus, the trapped water essentially includes an NaOH solution with a concentration approximated to 0.02 mol/L, rather than the concentrated carbonate/bicarbonate-based solutions associated with high-pH SCC. Moreover, the predicted potential is approximately 150 mV more negative than the lower end of the potential range where high-pH SCC occurs. The reason for this discrepancy is unclear in performed simulation and modeling. It was proposed [King et al., 2004] that conditions for high-pH SCC might be predicted over longer time periods (at least longer than a 30-day time period), particularly if the effects of CP are lost due to factors such as a high soil resistivity, to allow the potential of the pipe steel to enter the specific range for cracking to occur. However, a further model development and validation will be required to improve predictive accuracy.

Furthermore, in the field, precipitation of carbonate minerals occurs locally as a result of the increase in water pH. This process is considered in the modeled reaction mechanism, as seen in Figure 5-9 [King et al., 2004]. A consequence of the precipitation is decreased porosity of the coating. Within the model's 30-day simulation period, the greatest porosity reduction was predicted near the underside of the coating (i.e., that part closest to the pipe surface), and was caused by the precipitation of $FeCO_3$. Calcium and magnesium carbonate precipitations were also predicted to occur, but had a negligible effect on the coating's porosity. The precipitate first forms at the coating/soil interface and just within the coating and soil layers. With increasing time, the precipitate becomes more widespread, particularly extending into the soil layer. These predictions are consistent with field observations. Therefore, the modeling discrepancy in terms of high-pH SCC conditions is not attributed to the reduced coating porosity by inorganic precipitations.

Song model. Song [Song, 2008; Song, 2010] developed a model that predicts the solution chemistry and potential of pipe steel in a coating-disbonded region and their variations along the disbondment longitude with time under cyclic wet and dry soil conditions. It was claimed that some predicted results are consistent with field observations.

The model recognized the critical importance of coating conditions in the occurrence of high-pH SCC on pipelines. As shown in Figure 5-10 [Song, 2010], the following conditions would favor the SCC occurrence for disbonded coatings: (1) the presence of a holiday where the CP is applied; (2) the absence of holidays in the coating, but CP can permeate the coating; and (3) the absence of holidays in the coating, and the amount of dissolved oxygen in the trapped electrolyte is appreciable. The developed model targeted the scenario that the coating disbondment starts from a holiday, and the CP of -850 mV(CSE) is applied to the coated pipeline that is under cyclic wet-dry soil conditions. Figure 5-11 shows the distribution of potential vs. time beneath a disbonded coating. The potential adjacent to the open holiday (i.e., 0.001 cm) is almost identical to the CP potential applied. As the distance increases along the disbondment, the potential shifts less negatively. Upon removal of the CP, there is a much smaller change of the potential inside the disbondment than there is close to the holiday. This phenomenon has been attributed to CP shielding by the disbonded region (as previously analyzed). Moreover, when CP is lost during dry seasons, the potential inside the disbondment becomes less negative, and the potential distribution is more uniform across the disbonded region than it is during wet seasons when CP is enabled. It thus implies that the variation of solution chemistry during dry seasons is less significant than that during wet seasons.

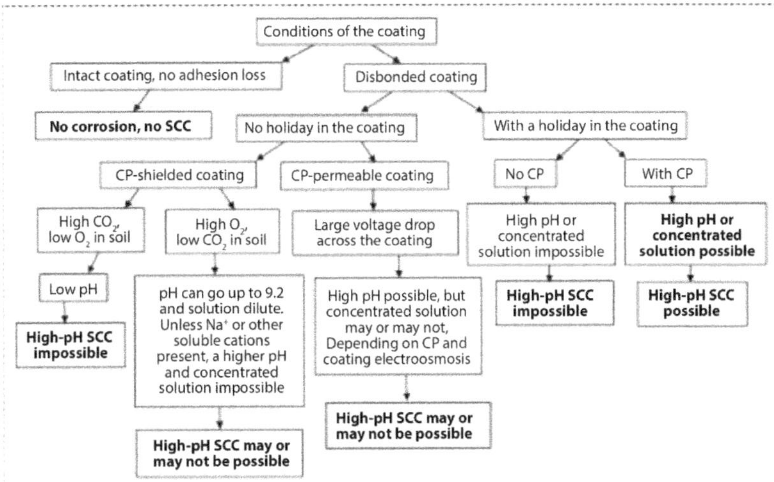

Figure 5-10. A schematic chart showing the different scenarios and conditions susceptible to high-pH SCC [Song, 2010].

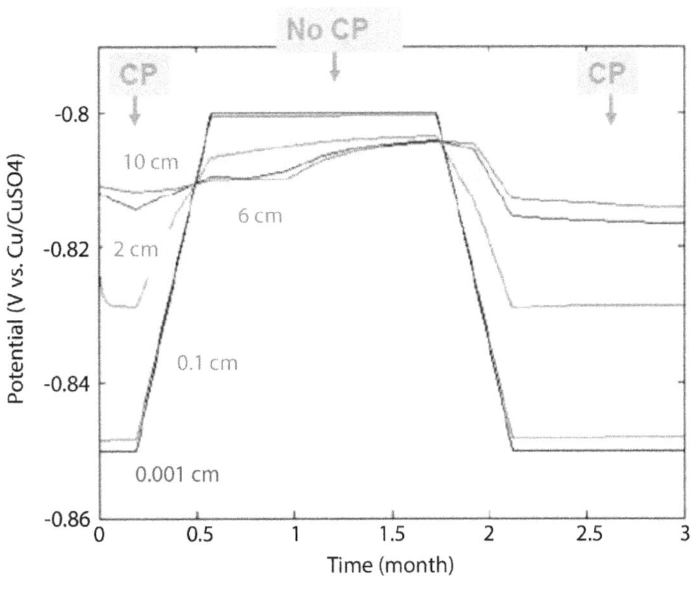

Figure 5-11. Distribution of potential vs. time under a disbonded coating during wet-dry cycling in soil [Song, 2008].

Figure 5-12 shows the variation in the pH of the trapped electrolyte beneath the disbonded coating as a function of time during wet-dry cycles. A sharp jump of pH occurs near the holiday, due to the generation of hydroxyl ions during CP-induced cathodic reactions. With an increased distance from the holiday, the pH increase becomes less significant due to the increased time required for the transport of solution species into the disbonded region (i.e., the shielding effect on CP penetration into the disbonding depth is increased). However, it is interesting to see that the pH inside the disbonded region is higher than it is at the holiday. The comment [Song, 2008] that the phenomenon is due to interactions of ionic transport and the electrochemical reactions on the steel surface is questionable. During dry seasons, CP is lost, which results in the drop of solution pH. This phenomenon can be observed during the long-term cycling of soil in wet-dry seasons.

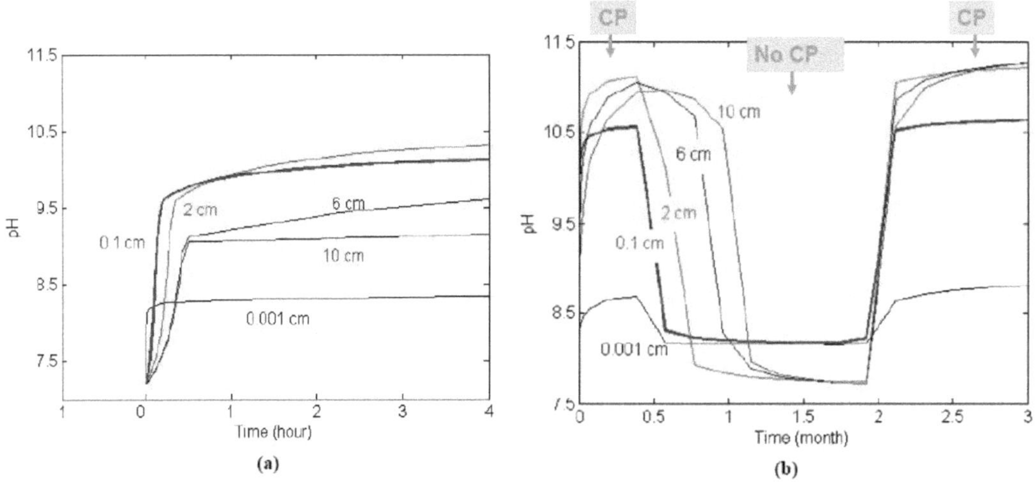

Figure 5-12. Variations of solution pH vs. time under disbonded coating during wet-dry cycling in soil [Song, 2008].

Figure 5-13 shows the profiles of HCO_3^-/CO_3^{2-} concentration in the disbondment over short- and long-term wet-dry cycling, respectively. The concentration of carbonate and bicarbonate ions increases with time. Over a long-term time period in the first wet season, the overall diffusion of carbonate and bicarbonate ions is dominated by bicarbonate ions, which diffuse into the disbonded region from the holiday due to a higher concentration of bicarbonate ions than carbonate ions. The overall concentration of carbonate and bicarbonate ions is determined by charge balance of the dominant species in the trapped solution (i.e., Na^+ and Cl^- ions). In the dry season, the solution becomes less alkaline and the concentration of carbonate and bicarbonate ions drops. In the second wet season, the concentration of carbonate and bicarbonate increases again. In the disbonded region, the carbonate and bicarbonate ions, rather than chloride ions, dominate the negative charge for high-pH SCC. It seemed that carbonate and bicarbonate ions repel Cl^- from the disbonded region when CP is effective.

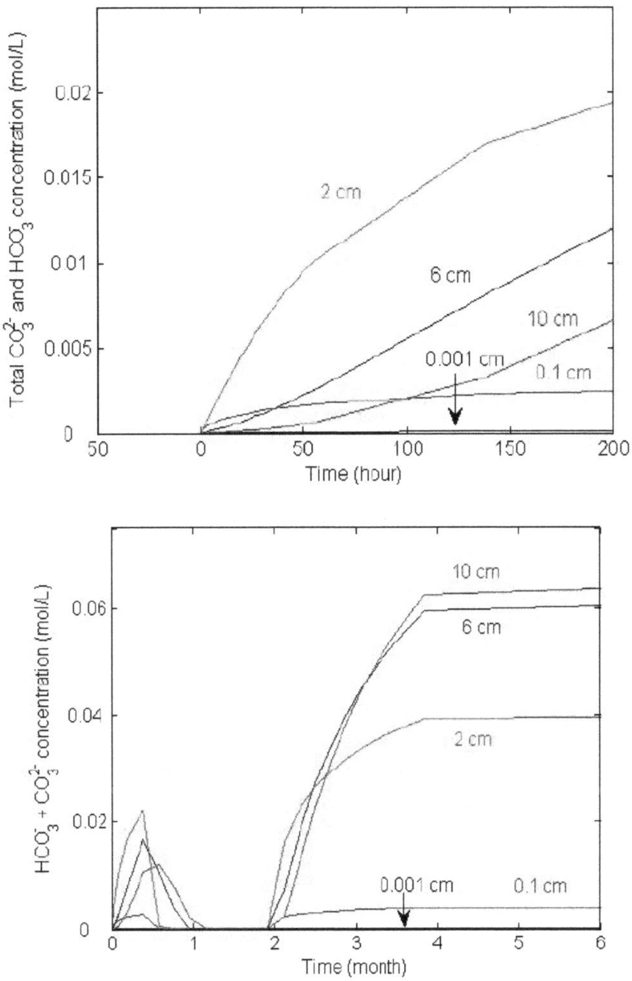

Figure 5-13. Profiles of HCO_3^-/CO_3^{2-} concentration in the disbondment over a short- and long-term of wet-dry cycling, respectively [Song, 2008].

The significant increase of the carbonate and bicarbonate concentration in the disbondment as modeled can be important for explaining why concentrated carbonate and bicarbonate ions are the dominant species in a high-pH SCC solution, despite being very diluted in the surrounding bulk-soil ground water. At even higher CP potentials, this concentration could be higher, while it is uncertain whether the ions could reach a point where sodium carbonate and bicarbonate solids will precipitate.

Furthermore (according to the model), the applied CP at a coating holiday could raise the solution pH and the concentrations of Na^+, CO_3^{2-}, and HCO_3^- in the trapped electrolyte beneath the disbonded coating. The solution chemistry in wet seasons and the soil's seasonal transitions could make the pipe steel be susceptible to high-pH SCC. Moreover, the CO_3^{2-} and HCO_3^- ions tend to displace more diffusive ions such as Cl^- from the coating disbondment. During a dry season, a high-pH solution may not be able to form without CP, even though a concentrated solution could form by water evaporation.

5.4.2. Near-neutral pH-solution Chemistry

The near-neutral pH-solution chemistry develops under disbonded impermeable coatings, such as PE tape or PE-based multi-layered coatings (e.g., 3LPE), which are incompatible with CP. The cracking environment associated with the near-neutral pH SCC features an anaerobic, diluted bicarbonate solution. Since the CP current is shielded from reaching the pipe steel surface, the steel at the SCC sites is at its corrosion potential of about -760 and -790 mV(CSE). Furthermore, there is no direct relationship between the occurrence of near-neutral pH SCC and the operating temperature of the pipeline.

Song model. Song et al. [Song et al., 2004] developed a mechanistic mode to predict the crevice chemistry initiating near-neutral pH SCC due to the degradation of a commercial HDPE coating mastic and the CO_2 penetration through the coating into the disbonded crevice. The model was supported by experimental testing results, including the degradation effects of the HDPE coating mastic on both the crevice solution pH and the dissolved CO_2 that permeates the coating. Mastic is a proprietary substance used with the HDPE coating that provides adhesion between the coating and the steel surface. The results show that degradation of the mastic used with HDPE coating decreases the water pH, which may help development of near-neutral pH environments leading to pipeline SCC. Moreover, in the field, the mastic degradation may contribute to the occurrence of near-neutral pH SCC by supporting microbiological activity [Jack et al., 1996] that results in environmental acidification. Furthermore, dissolved CO_2 penetrates the HDPE coating rapidly, equilibrating external environments in a few days. Depending on the partial pressure of CO_2 in the soil, the solution pH in coating a disbonded region can maintain near-neutral to support pipeline SCC.

A Tectran code simulates the crevice solution chemistry that is responsible for the occurrence of near-neutral pH SCC with an applied CP at the crevice holiday and continuous CO_2 permeation of the coating. A rectangular crevice with a holiday located at one end of the crevice was assumed. A general transport equation considered the diffusion and electromigration of relevant species and ions contained in the electrolyte trapped beneath the disbonded coating. Electroneutrality of charged species participating in corrosion reactions (both anodic and cathodic reactions) was combined to determine the distribution of various species along the crevice. The modeling predictions were compared to a number of experimental results, and an excellent agreement between the modeling results and the testing data was claimed. For example, Figure 5-14 shows the comparison of the simulated solution pH (separated from a bulk-soil solution where CO_2 is purged continuously) with the experimental results. A good agreement is shown when the coating permeability is 1.22×10^{-12} mol/atm m s, 10 times less than that reported in the literature (1.22×10^{-11} mol/atm m s) [Brandrup and Immergut, 1989]. This was measured when CO_2 was used as the gas phase. The model results greatly deviate from the experimental data when literature permeability was used.

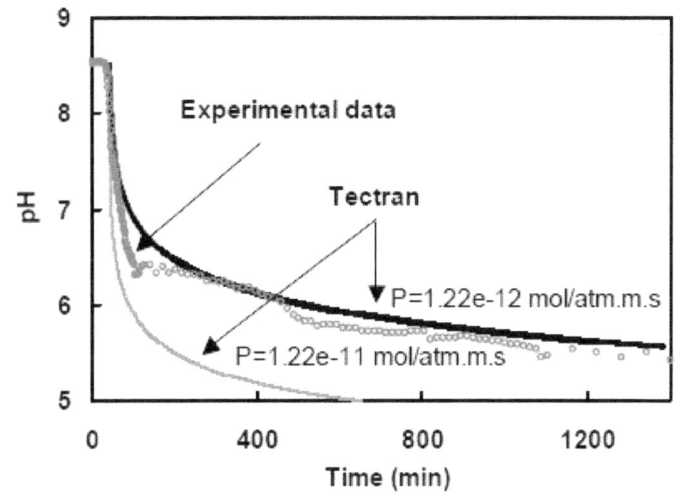

Figure 5-14. Simulation of the crevice pH using the Tectran code in comparison with experimental data. Two coating permeabilities were used for the simulation [Song et al., 2004].

Figures 5-15 shows pH values at the open holiday and the disbondment bottom, respectively, modeled by the Tectran code for the near-neutral pH environment. The initial pH of the solution prior to CO_2 purging is 8.0. When the solution is purged continuously with CO_2, the solution pH at the holiday is 6.4. The solution pH at the open holiday is elevated upon CP application at -0.97 V(CSE). Along with time, the pH values at both the holiday and the disbonding bottom increase due to the CP driving the cathodic reaction.

If the CO_2 penetrates the disbonding crevice only through the holiday, there is no obvious change of the solution pH at the disbondment bottom, at least within a short time period (since the CO_2 diffusion is slow and the length of the crevice is large). After 100 h, the penetration of CO_2 from the holiday has not yet reached the disbondment bottom. The steady pH of about 6.4 at the holiday indicates the quick transport of CO_2 to the holiday area from the bulk solution, which balances the increased pH resulting from alkali production by CP at the holiday. In addition to the holiday, the CO_2 can permeate the disbonding crevice through the coating film. The pH within the disbonding crevice decreases much more remarkably than it does at the holiday. Thus, the pH at the open holiday is dominated by bulk CO_2 chemistry, while the solution pH at the disbonding bottom depends heavily on CO_2 permeation of the coating film (since the diffusive pathway from the open holiday towards the crevice bottom is quite long). Furthermore, the solution pH at both the holiday and at the disbondment bottom are near-neutral due to the CO_2 penetration through the coating, even with the CP application at the holiday. The near-neutral pH environment that initiates pipeline SCC in the field may result from CO_2 permeation of the coating film.

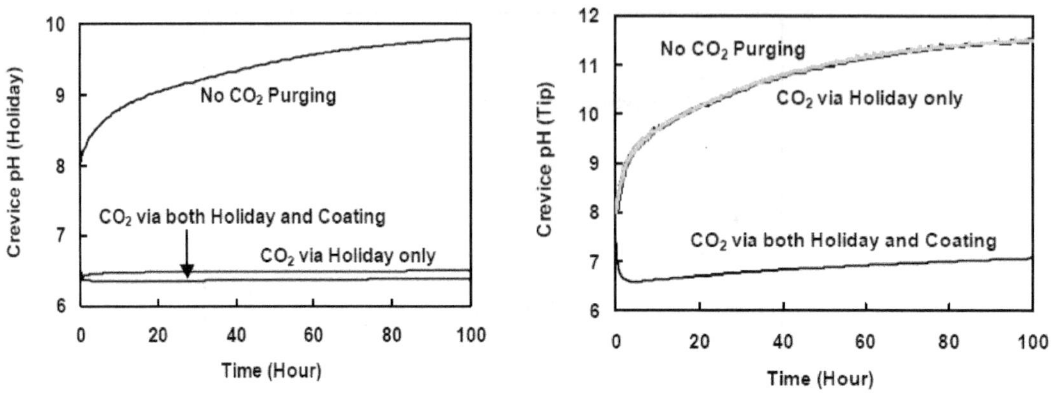

Figure 5-15. Simulated pH values at the holiday and the disbondment bottom, respectively, under three scenarios, indicating the effect of CO_2 penetrations through the holiday and through the coating film on the solution pH [Song et al., 2004].

Furthermore, Figures 5-16 and 5-17 show the variations of the solution pH and the amount of dissolved CO_2 with distance into the disbondment at different times when CO_2 can penetrate both the open holiday and the coating. Initially, the pH is 8.0 everywhere in the disbonding crevice and serves as an initial boundary condition. The solution pH at the holiday is kept at 6.3 as a boundary condition when time is longer than zero. The CO_2 diffuses from the holiday into the disbonding crevice and at the same time, permeates the coating film and reduces the pH of the solution trapped in the crevice. At 10 h, the pH reduces to about 7.0; and at 100 h, the solution pH nearly reaches the value of the soil solution. Furthermore, the total CO_2 concentration is 3×10^{-3} mol/L following the assumed initial condition. As time increases, the CO_2 diffuses into the crevice through both the holiday and the coating. The total CO_2 concentration increases with time. Within the crevice, the CO_2 transport through the coating film is dominant, with its concentration gradually built up over time. From the modeled results, the pH of the solution within the crevice was about 6.0–8.0, and the total CO_2 concentration was less than 2×10^{-2} mol/L. The consistence of the modeling results with experimental testing data indicates the significance of CO_2 penetration of the coating film, contributing to development of the near-neutral pH environments associated with pipeline SCC.

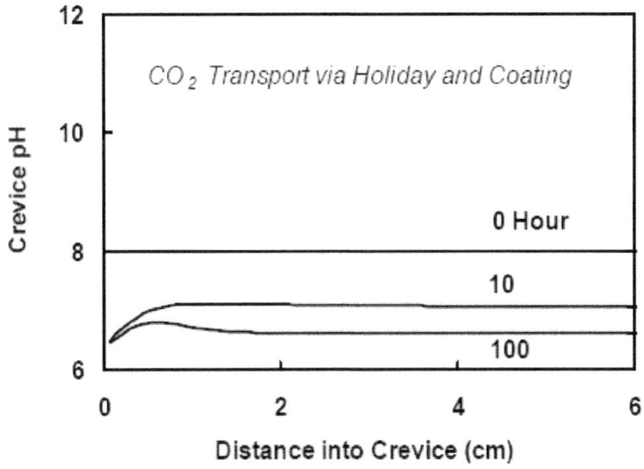

Figure 5-16. Variations of pH with distance into the disbonding crevice at different times for the scenario that CO_2 penetrates through both the open holiday and the coating film [Song et al., 2004].

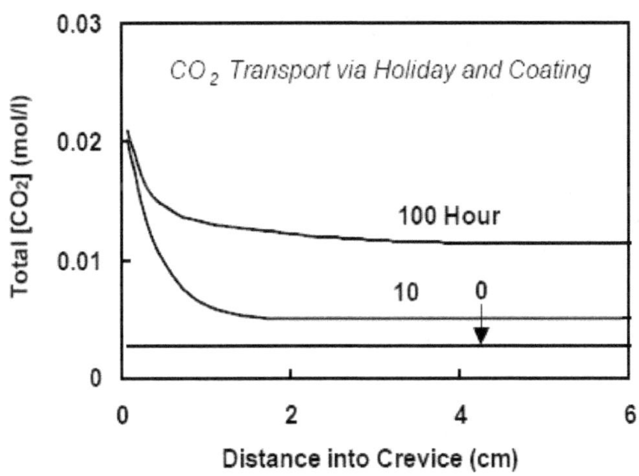

Figure 5-17. Variations of the total CO_2 concentration with distance into the disbonding crevice at different times for the scenario that CO_2 penetrates through both the holiday and the coating film [Song et al., 2004].

References

Beavers, J.A. (1992) *Assessment of the Effects of Surface Preparation and Coating on the Susceptibility of Line Pipe to Stress Corrosion Cracking*, PRCI report, L51666, USA.

Beavers, J.A., Thompson, N.G., Coulson, K.E.W. (1993a) Effects of surface preparation and coatings on SCC susceptibility of line pipe: phase 1–laboratory studies, *Corrosion'1993*, paper No. 93597, NACE, Houston, USA.

Beavers, J.A., Thompson, N.G., Coulson, K.E.W. (1993b) Effects of surface preparation and coatings on SCC susceptibility of line pipe: phase 2–field studies, *Proc. the 12th Inter. Conf. on Offshore Mechanics and Arctic Engineering*, ASME, USA.

Been, J. (2011) *Comparison of the Corrosivity of Dilbit and Conventional Crude*, Report no. 2480002, Alberta Innovates Technology Futures, Edmonton, AB, Canada, Sept. 2011.

Brandrup, J., Immergut, E.H. (1989) *Polymer Handbook*, 3rd Ed., John Wiley & Sons, Hoboken, NJ, USA.

Cheng, Y.F. (2007) Thermodynamically modeling the interactions of hydrogen, stress and anodic dissolution at crack-tip during near-neutral pH SCC in pipelines, *J. Mater. Sci.*, 42, 2701-2705.

Cheng, Y.F. (2010) Pipeline Engineering, in: Pipeline Engineering, Ed. Yufeng F. Cheng, in: *Encyclopedia of Life Support System*, Developed under the Auspices of the UNESCO, EOLSS Publishers, Oxford, UK.

Cheng, Y.F. (2013) *Stress Corrosion Cracking of Pipelines*, John Wiley Publishing, Hoboken, NJ, USA.

Cheng, Y.F., Niu, L. (2007) Mechanism for hydrogen evolution reaction on pipeline steel in near-neutral pH solution, *Electrochem. Commu.* 9, 558-562.

Fu, A.Q., Tang, X., Cheng, Y.F. (2009) Characterization of corrosion of X70 pipeline steel in thin electrolyte layer under disbonded coating by scanning Kelvin probe, *Corros. Sci.* 51, 186-190.

Heuer, J.K., Stubbins, J.F. (1999) An XPS characterization of $FeCO_3$ films from CO_2 corrosion, *Corros. Sci.* 41 (1999) 1231-1243.

Jack, T.R., Wilmott, M.J., Sutherby, R.L., Worthingham, R.G. (1996) External corrosion of line pipe a summary of research activities, *Mater. Perf.* 35, 18-24

Jones, R.H. (1992) *Stress Corrosion Cracking: Materials Performance and Evaluation*, ASM, Materials Park, Ohio, USA.

King, F., Jack, T., Chen, W., Wilmott, M., Fessler, R.R., Krist, K. (2000) Mechanistic studies of initiation and early stage crack growth for near-neutral pH SCC on pipelines, *Corrosion'2000*, paper no. 361, NACE, Houston, 2000.

King, F., Jack, T., Kolar, M., Worthingham, R. (2004) A permeable coating model for predicting the environment at the pipe surface under CP-compatible coatings, *Corrosion'2004*, paper no. 04158, NACE, Houston, TX, USA.

Li, M.C., Cheng, Y.F. (2007) Mechanistic investigation of hydrogen-enhanced anodic dissolution of X-70 pipe steel and its implication on near-neutral pH SCC of pipelines, *Electrochim. Acta* 52, 8111-8117.

Little, B.J., Lee, J.S., Ray, R.I. (2006) Diagnosing Microbiologically Influenced Corrosion: A State-of-the-Art Review, *Corros.* 62, 1006-1017.

Liu, Z.Y., Li, X.G., Cheng, Y.F. (2012a) Understand the occurrence of pitting corrosion of pipeline steel under cathodic polarization, *Electrochim. Acta* 60, 259-263.

Liu, Z.Y., Li, X.G., Cheng, Y.F. (2012b) Mechanistic aspect of near-neutral pH stress corrosion cracking of pipelines under cathodic polarization, *Corros. Sci.* 55, 54-60.

National Energy Board (1996) *Stress Corrosion Cracking on Canadian Oil and Gas Pipelines*, Report of the Inquiry, MH-2-95, Calgary, Canada.

Nishikata, A., Ichihara, Y., Tsuru, T. (1995) An application of electrochemical impedance spectroscopy to atmospheric corrosion study, *Corros. Sci.* 37, 897-911.

Parkins, R.N. (2000) A review of stress corrosion cracking of high-pressure gas pipelines, *Corrosion'2000*, paper no. 363, NACE, Houston, USA.

Parkins, R.N., O'Dell, C.S., Fessler, R.R. (1984) Factors affecting the potential of galvanostatically polarised pipeline steel in relation to SCC in CO_3^{2-}-HCO_3^- solutions, *Corros. Sci.* 24 (1984) 343-374.

Parkins, R.N., Blanchard Jr., W.K., Delanty, B.S. (1994) Transgranular stress corrosion cracking of high-pressure pipelines in contact with solutions of near neutral pH, *Corros.* 50, 394-408.

Perdomo, J.J., Chabica, M.E., Song, I. (2001) Chemical and electrochemical conditions on steel under disbonded coatings: the effect of previously corroded surfaces and wet and dry cycles, *Corros. Sci.* 43, 515-532.

Song, F. (2008) Overall mechanisms of high pH and near-neutral pH SCC, models for forecasting SCC susceptible locations, and simple algorithms for predicting high pH SCC crack growth rates, *Corrosion'2008*, paper no. 8129, NACE, Houston, TX, USA.

Song, F. (2010) Predicting the effect of soil seasonal change on stress corrosion cracking susceptibility of buried pipelines at high pH, *Corros.* 66, 095004-1 - 14.

Song, R.H., Pyun, S.I., Oriani, R.A. (1990) Hydrogen permeation through the passivation film on iron by time-lag method, *J. Electrochem. Soc.* 137, 1703-1706.

Song, F., Sridhar, N., Been, J., King, F. (2004) Predicting near-neutral pH SCC conditions under a disbonded coating on pipelines, *Proc. Inter. Pipeline Conf. 2004*, IPC04-0196, Calgary, AB, Canada.

Stratmann, M., Streckel, H. (1990a) On the atmospheric corrosion of metals which are covered with thin electrolyte layers - I. Verification of the experimental technique, *Corros. Sci.* 30, 681-696.

Stratmann, M., Streckel, H. (1990b) On the atmospheric corrosion of metals which are covered with thin electrolyte layers - II. Experimental results, *Corros. Sci.* 30, 697-714.

Stratmann, M., Streckel, H., Kim, K.T., Crockett, S. (1990) On the atmospheric corrosion of metals which are covered with thin electrolyte layers - III. the measurement of polarisation curves on metal surfaces which are covered by thin electrolyte layers, *Corros. Sci.*, 30, 715-734.

Tang, X., Cheng, Y.F. (2011) Quantitative characterization by micro-electrochemical measurements of the synergism of hydrogen, stress and dissolution on near-neutral pH stress corrosion cracking of pipelines, *Corros. Sci.* 53, 2927-2933.

Van Boven, G., Chen, W., Rogge, R. (2007) The role of residual stress in neutral pH stress corrosion cracking of pipeline steels. Part I: Pitting and cracking occurrence, *Acta Mater.* 55, 29-42.

Xu, L.Y., Cheng, Y.F. (2013) Development of a finite element model for simulation and prediction of mechano-electrochemical effect of pipeline corrosion, *Corros. Sci.* 73, 150-160.

Yu, J.G., Luo, J.L., Norton, P.R. (2001) Effects of hydrogen on the electronic properties and stability of the passive films on iron, *Appl. Surf. Sci.* 177, 129-138.

Zhang, G.A., Cheng, Y.F. (2010) Micro-electrochemical characterization of corrosion of pre-cracked X70 pipeline steel in a concentrated carbonate/bicarbonate solution, *Corros. Sci.* 52, 960-968.

Pipeline Coating Performance Testing

6.1. Introduction

Testing a coating system to determine how it will perform in a certain environment is a critical part of coating selection. Many industry standards exist for testing coatings. NACE International has many well-written standards that are a good start. The International Organization for Standardization (ISO), ASTM, AWWA, CSA, and many countries have different standards. Without understanding how a coating system will perform in a particular environment, the end user is taking a chance of not knowing how a coating will perform.

The only true test is to bury the applied coating in the environment (ground or water) and expose it at different times throughout its life to see how it is performing. This is not practical, since companies, engineers, and coating manufacturers need some guidance about whether a coating system will or will not perform as required in a particular environment. In some cases, it may take several years before coating system failures are evident. Many coating systems looked good on the pipes and passed required tests only to fail after a few years of service.

The failure mode of a coating is critical with CP. If the coating adhesion fails (disbonds) and water penetrates the disbondment, is CP allowed to protect the pipe or not? When selecting a pipeline coating, the "Fail Safe" (or non-shielding) characteristics may be more important than other issues that are normally considered. [Norsworthy, 2006]

To determine if a candidate coating system may provide satisfactory service under the variable conditions anticipated during coating application, installation, and service, an attempt should be made to replicate the conditions in the laboratory, accelerate them if possible (to provide for differentiation between candidate systems), and to establish a ranking prioritization between candidate systems. [Tator, 2006]

Figure 6-1. Cathodic disbondment test (ASTM G-8) shows growth of the disbondment area from 30 to 60 to 90 days.

Most laboratory testing is short-term and therefore gives short-term results. This does not keep an engineer from requiring a longer-than-standard test to understand if a coating will continue to be affected by the testing. There are times when a coating system passes a 30-day test, but at 60 days and 90 days there is growth of a failure. Continued growth of any failure is not acceptable. This is shown in the photos of cathodic disbondment test results in Figure 6-1.

The longer the test, the more reliable the results. When a testing regime is designed, time is an important part of the program and should be well-planned. The type and number of tests are also important. Completing a variety of tests on the same coating provides a thorough understanding of the coating's properties. No one test provides enough information for proper coating selection.

One must also be willing to test beyond traditional industry testing. However, many standard tests are designed for specific environments that may not be entirely appropriate, so variations may need to be employed. [Al-Borno, 2005] If industry tests are modified for a particular environment, the modification must be well-documented. When using industry standard testing to compare various coating systems side-by-side, run the tests per the standard to ensure accurate information for each type of system. If testing for a particular environment, then one can adapt standard tests to meet the environmental needs.

Typically, to qualify and compare coating types for a particular environment, the tests are performed for longer timeframes. Production testing is usually shorter in length than coating qualification testing.

Many industry standards can be used to find descriptions of various coating tests. Each industry, organization, and country may have slightly different requirements for testing, but test requirements are generally similar. When testing coating, it is critical to simulate the actual environment in which the coating will be in service. Of course, it is not possible to do this in all cases. Table 6-1 summarizes the main standards developed by different organizations and countries for coating tests.

Table 6-1. List of Standards for External Coating Materials and Application

Standard Designation	Title
ANSI/AWWA C 203	Coal-Tar Protective Coatings and Linings for Steel Water Pipelines—Enamel and Tape—Hot Applied
ANSI/AWWA C 213	Fusion-Bonded Epoxy Coating for the Interior and Exterior of Steel Water Pipelines
ANSI/AWWA C 214	Tape Coating Systems for the Exterior of Steel Water Pipelines
ANSI/AWWA C 215	Extruded Polyolefin Coatings for the Exterior of Steel Water Pipelines
ANSI/AWWA C229	Fusion-Bonded Polyethylene Coatings for Steel Water Pipe and Fittings
ANSI/AWWA C2GT	Geotextile Backed Cold-Applied Tape Coatings for Steel Water Pipe, Special Sections, Connections and Fittings
CSA Z245.20/Z245.21	External Fusion-Bond Epoxy Coating for Steel Pipe/External Polyethylene Coating for Pipe
DIN 30670	Polyethylene Coatings of Steel Pipes and Fittings; Requirements and Testing
NACE No. 12/AWS C2.23M/ SSPC-CS 23.00	Specification for the Application of Thermal Spray Coatings (Metallizing) of Aluminum, Zinc, and Their Alloys and Composites for the Corrosion Protection of Steel
NACE Standard SP0185	Extruded Polyolefin Resin Coating Systems with Soft Adhesives for Underground or Submerged Pipe
NACE Standard SP0394	Application, Performance, and Quality Control of Plant-Applied, Fusion-Bonded Epoxy External Pipe Coating
NACE Standard RP0399	Plant-Applied, External Coal-Tar Enamel Pipe Coating Systems: Application, Performance, and Quality Control
NACE Standard RP0402	Field-Applied Fusion-Bonded Epoxy (FBE) Pipe Coating Systems for Girth Weld Joints: Application, Performance, and Quality Control
NACE Standard RP0602	Field-Applied Coal Tar Enamel Pipe Coating Systems: Application, Performance, and Quality Control
NACE Standard TM0204	Exterior Protective Coatings for Seawater Immersion Service
NACE Standard TM0304	Offshore Platform Atmospheric and Splash Zone Maintenance Coating System Evaluation
NACE Standard TM0404	Offshore Platform Atmospheric and Splash Zone New Construction Coating System Evaluation
NACE RPO	Field Applied Two part epoxies
NACE RPO	Field Applied Shrink Sleeves
ISO 21809-2: 2007	Petroleum and natural gas industries-External coatings for buried or submerged pipelines used in pipeline transportation systems-Part 2: Fusion-bonded epoxy coatings
NACE SP0109-2009	Field Applied Tape Coatings

In addition to inspections conducted in the plant and the field, other quality-control testing is performed in a test laboratory. Most coating facilities have quality-control laboratories to ensure the quality of the production coating process. Generally, laboratory testing is performed on coating materials to determine the coating type or manufacturer. This allows the end user to separate coating systems that do not perform in a particular environment from those that do. Even though these are short-time tests, they provide valuable information for the selection process. Each test provides a different set of data for selection criteria. The quality of test data contains critical information for the end user.

Some of the most frequently performed tests include:

- Cathodic Disbondment: All pipeline coating types used with CP
- Hot Water Adhesion: All pipeline coating types
- Flexibility: All pipeline coating types where pipe is bent or flexed in field
- Porosity (Cross Section and Interface): FBE and liquid coating types
- Gel time: FBE coatings only
- Adhesion: All pipeline coating types
- Impact Resistance: All pipeline coating types
- Thermal Characteristics: FBE and liquid coating types

6.2. Cathodic Disbondment

Generally, this test assesses the resistance of coatings bonded to the metal substrate when they are exposed to CP. These may be long-term or short-term tests, according to whether the testing is for coating selection or production testing.

6.2.1. Testing Standards

ASTM G-8. ASTM G-8 is popular for testing coating types that can be easily applied to small-diameter pipes (2" to 4", or 5 cm to 10 cm). Once cured, three intentional holidays are drilled through the coating to expose the substrate. The size of the holidays usually ranges from 0.125" to 0.250" (3.3 mm to 6.6 mm) in diameter. One end of the pipe is capped and sealed so no water or current will enter the pipe at that end. A hole is drilled through each side of the pipe top for dowel placement (to support the pipe in a vertical position during the test duration).

These pipes are immersed in a mixture of 3% salt water. The consistency of the salt can vary per the standard or test method used. The electrolyte is kept at 72°F (25°C). The pipes are immersed until all drilled holidays are completely covered by the electrolyte. Only the coated parts of the pipe and holidays should be in the electrolyte (because any uncoated pipe will consume the CP current).

Once the pipes (there can be several in the same tank) are immersed, the pipes are connected to the negative terminal of the DC power supply. The impressed current anode is attached to the positive side of the power supply. This is critical. If these leads are reversed, the coated sample will become the anode and corrode rapidly at the intentional holidays. There can be a variety of anode materials used, but the high silicon cast iron rods or mixed metal rods are popular.

The power supply is energized and potentials are set to test parameter values. The reference electrode for determining the correct potential setting for the test is either a calomel (mercury-based) or silver/silver chloride. These references are both used for taking potentials in salt-water environments and are not contaminated by the chlorides in the electrolyte. Since calomel electrodes are mercury-based, proper disposal is critical and for this reason some labs are converting to silver/silver chloride.

The position of the reference cell in relationship to the test sample is important. It should be placed where the particular standard states when the potential is taken. The DC power supply should have individual adjustments so each sample's potential can be adjusted to the proper value. These potentials should be measured and adjusted as needed each day, since polarization and other electrochemical reactions will change their values.

These tests are typically run for 30, 60, or 90 days, but can be extended for longer-term information. Monitoring of potential, electrolyte temperature, and adjustment data should be recorded daily or at least on each day the laboratory is open.

Evaluation methods of cathodic disbondment tests will be discussed in detail later. In some testing programs, one of the three holidays is tested at 30 days, another at 60 days, and the last after 90 days. Figure 6-2 shows the experimental setup of the attached cell method for cathodic disbondment testing.

Figure 6-2. Attached cell method of cathodic disbondment testing.

ASTM G-42. ASTM G-42 is the same as the ASTM G-8, except that the test is run at elevated temperatures. Each temperature is determined by the environment in which the pipe will be placed when in service. In some tests, the electrolyte is heated. In other tests, the pipe is heated with circulating heat transfer oil through the pipes. Other test methods have been used for heating the pipe.

ASTM G-95. ASTM G-95 describes the attached cell method used for testing many types of coating systems, especially production samples in an FBE plant. This test method can be run at ambient temperature or higher temperatures. Coating samples may be cut from the production pipe, or they can be square, precut plates that have been coated per the specification.

This can be long-term or short-term testing. If the test is run at elevated temperatures, the samples are placed on large, flat hot plates. There is usually a heat transfer material between the plate and the samples. The material may be metal shot, grit, or sand, and provides a more consistent heat to the panels.

The intentional holiday is drilled through the coating. A plastic cylinder is centered over the plate and holiday, then chalked with a silicon material to seal the cylinder to the plate. The diameter and height of the cylinder is determined by the particular test procedure, the size of the sample, and the amount of water needed. The plastic cylinder can be clear or colored, but it must be able to withstand the temperatures of the test and hold the proper amount of electrolyte. Typical size is a 4" (10 cm) diameter by a 6" (15 cm) height.

After the silicon is dry, a 3% salt water solution is placed in the cylinder, which is set on the hot plate and allowed to heat to the test temperature. According to the test method, either the electrolyte temperature is used or the plate temperature is used to control the test.

The anode is placed in the electrolyte and attached to the positive terminal of the power supply. The negative pole of the power supply is connected to the plate. To ensure that a proper potential is set, a multi-meter is used with either a calomel or silver/silver chloride reference cell connected to the negative terminal of the multi-meter and the positive terminal is connected to the sample plate.

The power supply should have a separate adjustment for each test sample. The proper voltage is set in each separate sample per test requirements. The voltage setting is determined by the test procedure. Typically, long-term tests use a 1.5 V setting (as compared to the reference). Short-term tests are usually set at a higher voltage (such as 3.0 V or higher). The temperature and voltage used in the test can be modified as needed to simulate certain conditions or to accelerate the test. Higher voltages generate more hydrogen gas than lower voltages and may accelerate the disbondment. The higher temperatures increase the electrochemical process of alkalinity in the cell in an effort to accelerate the disbondment.

6.2.2. Testing Evaluation

Using a utility knife or another tool, make radial cuts at about 45° from the edge of the holiday outward through the coating to the metal. The radial cuts must be at least 0.8 in. (20 mm) in length. Insert the blade of the knife under the coating at the edge of the holiday at the point of each cut.

Use a prying action to chip off the coating. Continue this process until the coating demonstrates a definite resistance to the prying action. The radius of the disbonded area from the holiday edge along each radial cut is measured and the results are averaged and recorded along with other pertinent information about the coating and the test parameters.

To properly evaluate a cathodic disbonding test, one must understand what the test does. Cathodic disbondment is typically a circular pattern extending out from the edges of the holiday. If true cathodic disbondment has occurred, the coating will easily be removed in this circular pattern, because the coating has lifted from the substrate due to electrochemical reactions. Little force is needed to remove most disbonded coatings. There should be no coating left in the profile of the disbondment area.

With hard coatings (such as FBE and two-part epoxies), the thickness of the coating may make it difficult to cut through and remove the disbonded area. In this case, a rotary tool or other devices can be used to make "pie cuts" around the holiday. These cuts must go to the metal substrate. Lifting the disbonded area is also more difficult, since the coating is thick and may require considerable force to remove it from the surface. Figure 6-3 shows the evaluation of an attached cell cathodic disbonding test on FBE coating.

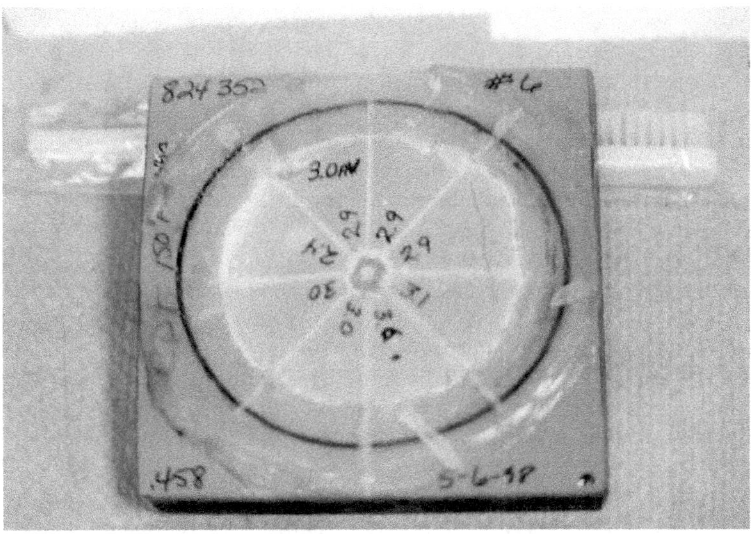

Figure 6-3. Evaluation of an attached cell cathodic disbonding test on FBE coating.

The knife should not follow the pie-cut lines. Cutting the lines can crack and loosen some coating types, but this is not due to cathodic disbondment.

Softer coatings such as tapes, shrink sleeves, and wax coatings are more easily cut and removed. The same is true with the cathodic disbondment area on these coatings, which makes coating removal easy. The disbonding area will typically be circular. Some evaluators have made significant mistakes by using pliers or other such devices to clamp onto the coatings and trying to remove the coating with force, which is not cathodic disbondment.

There are times when certain coatings will lose disbondment because of water penetration and the coating will easily lift past the cathodic disbondment area. This should not be evaluated as cathodic disbondment, but as the loss of adhesion from water penetration. This can be recognized with the discoloration of the actual cathodic disbondment ring. Anything past this point is not considered to be cathodic disbondment. Many times the coating beyond the cathodic disbonding ring will actually re-attach itself to the substrate when the test sample is left to dry. Also, the coating outside of the test cylinder will be adhered. Any coating left over the ring will never adhere to the substrate as, in the case of the loss of adhesion from water penetration.

6.3. Hot Water Adhesion

This test provides information about how well a coating can withstand water penetration. The water is heated to accelerate its penetration of the coating. Generally, pure water is a universal solvent and will penetrate a coating faster than water with contaminants.

This test can be run similarly to the cell-attached cathodic disbonding test, but the test sample does not have a holiday. After the cylinder is attached and the caulk is dry, the heated distilled water is added and the cell is set on a hot plate. The temperature is set to the required level, typically 150° F (65°C), and maintained within 5°F (3°C) throughout the test period. Again, the temperature can be higher or lower if needed to meet the anticipated environment.

The other method is to completely immerse the coated sample in a hot-water bath. The temperature range must be maintained throughout the test period. In this case, the water can penetrate the coating at the edges of the sample. Sometimes, samples are sealed on the edges to keep water from penetrating the coating edges.

Evaluation is performed by cutting through the coating to the substrate in a rectangular pattern. Each corner is checked for adhesion loss. A utility knife is used as a lever to lift the coating at each corner of a hard coating. NACE SP0394 provides guidance for grading the samples according to how easily and how much coating is removed in the tested area. Some guidelines have testers making an "X" cut and evaluating that area to determine the level of adhesion loss. Softer coatings are removed with a type of pliers or clamping device: The coating is easily pulled off without cohesive failure if true adhesion loss has occurred.

A coating with true adhesion loss is easily removed with little or no coating remaining in the profile. Any coating can be cut from the surface with enough time and effort, but adhesion loss causes it to be removed easily with little effort. Each coating type has to be evaluated differently.

6.4. Flexibility

Flexibility of the applied coating is important for determining whether the coating will crack during construction handling, field bending, and other possible stresses. Nearly all coatings can be tested for flexibility.

For FBE, curing results in reduced flexibility, so the test provides a good indication of cure. Foam bond and a contaminated substrate may also negatively affect flexibility of all coatings, and this test provides insight into these problems if they exist.

There are two methods for testing flexibility (i.e., bending mandrels of fixed radii or four-point bending apparatus). Test specimens are cut approximately 1" X 8" (2.5 X 20 cm) from the pipe with the 8" (20 cm) dimension parallel to the length of the pipe, as shown in Figure 6-4. Use a file to remove all burrs; smooth the edges of the bar and coating.

Figure 6-4. Facility and the specimens for testing of flexibility of coatings.

If the test temperature is 32°F (0°C), specimens are chilled below the freezing point in a freezer or with dry ice, and then allowed to warm. When frost on the specimen melts, the temperature is in the correct range. There are times when the test temperature is lower than 32°F (0°C) because the coated pipe will be handled and bent in much colder temperatures. The expected most extreme temperature is used in these cases to ensure the coating will not crack during construction in cold environments such as -22°F (-30°C).

When the test panels reach the correct temperature, position them in the bend apparatus and start bending. Bending should be completed within 30 seconds. As test panels are being bent, look for stretch marks or cracks in the coating. If you see cracks during bending, the test can be stopped. Any cracks or tears in the coating is considered a failure. Those within 1/8" (3 mm) of the strap edge should be ignored, since they may be caused by stress.

6.5. Porosity and Interface Contaminants

Porosity at the coating and substrate interface can cause lack of adhesion. Porosity within the coating can allow water to more easily penetrate the coating. Porosity also indicates that there are problems with the application process or handling of the powder, such as increased moisture levels in the plant-coating systems. Moisture can be absorbed by the powder during the recycling process. Moisture must be removed using compressed air with a dew point of -20°F (29°C) at 60 psi or drier in the fluidizing and powder transport systems. Application temperatures that are too high can also cause porosity.

Abrasive cleaning always leaves residue on the cleaned surface. High residue levels may reduce a coating's long-term durability and installation characteristics, therefore indicating that improvements in the cleaning process are in order. This test is useful only for contaminants visible under 30–40X magnifications.

Contaminants such as inorganic salts or organic oils, which may be detrimental to coating performance, may not be detected by this method. Due to the severity of this test, even properly applied coatings will separate from a well-prepared substrate. A visual examination is necessary to determine the level of contamination.

The bend test can create samples for determining if porosity is present at the interface or within the coating itself. By getting the sample extremely cold and bending it rapidly, the FBE or other coating will pop off the metal surface. These chips are then looked at under a microscope to determine the size and extent of the interface and cross-section porosity. The bend test also provides an indication of the amount of visible residue left on the metallic substrate by the abrasive cleaning process.

NACE SP0394 provides guidance for grading porosity and what is acceptable.

6.6. Gel Time

This test determines how much time is required at a certain temperature for the heated powder to transform from a liquid to a gel or semi-solid. Advanced powder can cause significant reduction in gel time, and may have an orange peel appearance, poor flow out, or a dull, sandy finish in the applied coating. The test is subjective, and to obtain reproducible results, the testing technician should be consistent with the test performance. Temperature is critical, since a 10°F (6°C) change in temperature can result in a 20% change in gel time.

During testing, a small amount of FBE powder (1 gram or less) is applied to a hot plate while maintaining at 400 ± 2°F (204 ± 1°C). The hot-plate surface must be clean and preheated before the powder is placed on it. Check plate temperature with a surface thermometer to ensure the correct temperature range. After the powder is placed on the pre-heated plate, start the timing device. Begin stirring the powder with a stiff wire until it can no longer be stirred. Another method is to drag a spatula across the powder until it rides up the spatula blade, showing the powder has gelled as shown in Figure 6-5. Stop timing and record the results. The test should be run three times and the results averaged. If the test completes outside of the parameters specified, the powder should be rejected.

Figure 6-5. FBE powder gel test.

6.7. Impact Resistance

This test method provides information about how well a particular coating will resist damage from impacts during handling, transportation, and construction. An ASTM G14 Impact Tester or equivalent is used. The coated sample can be of different sizes and shapes according the type of coating being tested.

A tup is a heavy metal object that can accommodate a 0.625" (15.9 mm) diameter ball bearing. The tup shall have a hardness of 50 to 55 HRC. The apparatus including the ball bearing at the bottom typically weighs 2.2 lbs (1.0 kg). providing an adequate impact energy when it is dropped.

The base of the impact tester is typically a square block of laminated wood (some standards are very specific about the type of wood) that is 24" (610 mm) by 24" (610 mm). This block is usually topped by a hardwood layer. The impact tester is firmly screwed to the wood block. The slotted cylinder holding the weighted apparatus with the ball bearing is typically 39" (1.0 m) in height and has gradual measurements. These measurements may be in inches or centimeters as needed for the particular standard. A flat steel anvil is placed on the wooden base and the flat coated sample is placed on the top of the anvil and held in place with clamps. Test ring samples are placed in the mandrel to hold them in place during the test. These can be modified per the size and shape of the coated test specimen.

The weighted apparatus is lifted to a known height and dropped onto the coated surface. The number of times the weight is dropped and the height of the drop will be determined by the particular test method and type of coating being tested.

A holiday detector is set to an appropriate setting per the test method and type of coating being tested. After each impact, the test sample is checked with the holiday detector to determine if the impact damaged the coating enough to cause a holiday.

The maximum amount of energy (J or in-lb) that the coating absorbed without exposing a holiday is calculated and recorded with other pertinent information as required by the particular test method. Figure 6-6 shows the setup for testing a coating's impact resistance.

Figure 6-6. Setup for testing a coating's impact resistance.

6.8. Glass Transition and Heat of Reaction Determination

This test provides information about the glass transition temperature and the amount of exothermic heat of reaction for certain types of coating. The tests are used on FBE and other coatings to help determine the amount of cure for the coating.

A differential scanning calorimeter (DSC) machine is used for the tests. A small amount (10 ± 1 mg) of the cured coating is placed in a pre-weighed aluminum pan, which has a crimped cover. Create a small vent hole in the lid without damaging where the pan is placed in the DSC machine. Perform the test as described in the specific procedure. This test procedure is complicated and should only be performed by trained, experienced technicians.

The CD performance of both conventional and new generation FBE coatings improves with an increase of cure percentage as determined by differential scanning calorimetry (DSC). [Zhou, Edmondson, Jeffers, 2006]

The scans developed during the test provide the glass transition measurement (when the coating starts to deteriorate due to heat). The residual exothermic heat of reaction can be calculated following the instructions provided by the manufacturer of the DSC equipment.

DSC can also be used in coating forensics to help determine the cure of the existing coating, which may provide information needed to determine coating failure, etc.

6.9. CP Shielding Tests

With all the misunderstanding about disbonded coating and its potential for shielding CP, the pipeline industry has begun developing laboratory testing methods for determining if a coating is CP-shielding or not when the coating disbonds and water penetrates beneath the coating.

This topic has been discussed in previous chapters. These references provide guidance for future testing of coatings to determine if a particular coating is shielding or non-shielding to CP.

Gaz de France and Total have carried out lab studies to assess the influence of the main parameters governing the corrosion rates beneath a simulated coating disbondment as a function of the distance from the point a direct contact exists with the external electrolyte, at the end of the disbondment. [Roche, 2007

In 2014, NACE International started a task group to develop test methods for determining if a coating is shielding or non-shielding to CP when disbondments occur. The objective of Task Group 521 is to assess the "shielding" characteristics of pipeline coatings with cathodic protection. [Al-Borno, 2014]

Can the vendor provide proof of non-shielding to CP when disbondments occur through third-party laboratory testing and years of proof in the field? [Norsworthy, 2013] Each operator has to know their own system before they can be certain if CP potentials will penetrate beneath their disbonded coatings. [Jack, 1994]

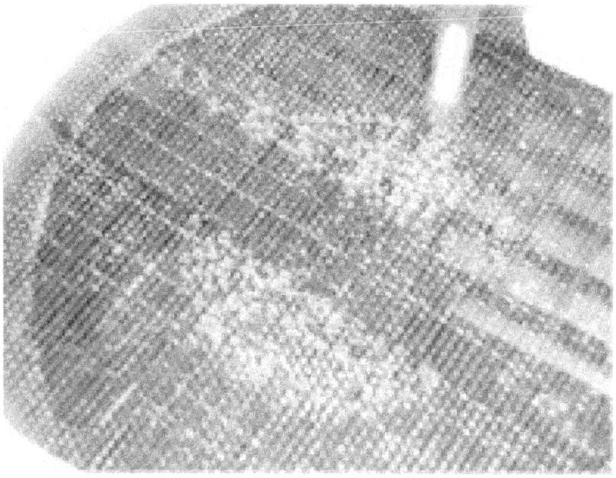

Figure 6-7. Testing for non-shielding properties of mesh-backed tape show hydrogen evolution.

Dr. Cheng and many others have authored related papers on this topic. Many of these are listed in other chapters' references.

REFERENCES:

Al-Borno, Amal; Brown, Mick; "Testing and Selection of Girth Weld Protective Coatings," CORROSION 2005, Paper 05031

Al-Borno, Amal; "Laboratory Test Methods for Evaluating Cathodic Current Shielding," Task Group 521 Presentation, 2014

T.R. Jack, G. Van Boven, M. Wilmott, R.L. Sutherby and R.G. Worthingham; "Cathodic Protection Potential Penetration under Disbonded Pipeline Coating," Materials Performance, August 1994

Norsworthy, R., "Is Your Pipeline Coating 'Fail Safe'?" Pipeline and Gas Journal, October 2006, pg. 62, Volume No. 233 Number 10.

Norsworthy, R., "Understanding Mesh-backed Coating System and Non-shielding," Corrosion 2013, Paper 2205

Roche, M. "Corrosion Management: A Key issue in Pipeline Integrity," International Petroleum

Ruschau, G., Digulio, J., "Cathodic Protection Shielding by Pipeline Coatings," AMI 2016 paper

Tator, Kenneth B., "Laboratory Testing of Pipeline Exterior Coatings-a commentary on correlation with field performance," CORROSION 2006, Paper 06042

Wenjing Zhou, Stephen J. Edmondson and Thomas E. Jeffers, "Effects of Application Temperature, Degree of Cure and Film Thickness on Cathodic Disbondment of Conventional and New Generation FBE Coatings," CORROSION 2006, Paper 06049

7 Coating Evaluation by Electrochemical Techniques

Corrosion of metals is electrochemical in nature [Stansbury and Buchanan, 2000]. Corrosion scientists and engineers have used various electrochemical measurement techniques to investigate corrosion phenomena, including corrosion of metal beneath coatings or at coating defects, to obtain mechanistic and kinetic information about the corrosion process. These electrochemical techniques, such as those included in this chapter (i.e., electrochemical impedance spectroscopy, localized electrochemical impedance spectroscopy (LEIS) and scanning Kelvin probe) are either quite convenient to use [Frankel, 2008] or can provide unique corrosion information at a microscopic scale that is unavailable by conventional techniques as measured by a macroscopic electrode [Cheng, 2011]. Moreover, compared to the classic weight-loss method for obtaining corrosion rate, the electrochemical techniques are quite time-saving.

This chapter introduces three types of advanced electrochemical techniques (i.e., EIS, LEIS and SKP), which have been used to study coatings (including pipeline coatings) and the corrosion of coated metal specimens. It is anticipated that these techniques enable investigation of corrosion fundamentals beneath coatings and the evaluation of coating performance, as well as the development of high-performance coating technology for improved pipeline integrity.

7.1. Electrochemical Impedance Spectroscopy

Electrochemical impedance spectroscopy (EIS) is a powerful technique for characterization of electrochemical systems, including corrosion processes. One of the most promising features of EIS is that, with a sufficiently broad range of measuring frequencies, the influence of governing physical and chemical steps in the whole electrochemical process can be distinguished and measured. For corrosion processes, the individual steps (such as charge-transfer, mass-transfer, film-formation, etc.) can be recorded in the different frequency ranges of the impedance spectroscopy, and are thus studied separately to avoid any effects from other steps.

Currently, EIS has been routinely used to characterize coatings and the study of corrosion beneath the coating. One of the most successful applications of EIS is in the evaluation of polymeric-coating properties and the corrosion of coated metals, as well as changes to coatings during exposure to corrosive environments [Mansfeld, 1995]. Measured EIS data gives both mechanistic and quantitative information. Data include various parameters associated with charge-transfer reactions and diffusive steps occurring at the metal/coating and coating/electrolyte interfaces and in the coating matrix and solution boundary layer, respectively. Quantitative information includes coating resistance and capacitance, charge-transfer resistance, double-charge layer capacitance, etc.

7.1.1. The Technique and Measuring Principle

The fundamental principle of the electrochemical impedance approach is to apply an alternating-current potential signal to the system under investigation, and to measure the current response. The applied potential signal is usually a sinusoidal excitation signal with a small amplitude. The response to the applied potential signal is an alternating-current signal. As the applied signal is sufficiently small, the system response is pseudo-linear, where the current response to the sinusoidal potential will be a sinusoid at the same frequency but shifted in a certain phase, as shown in Figure 7-1.

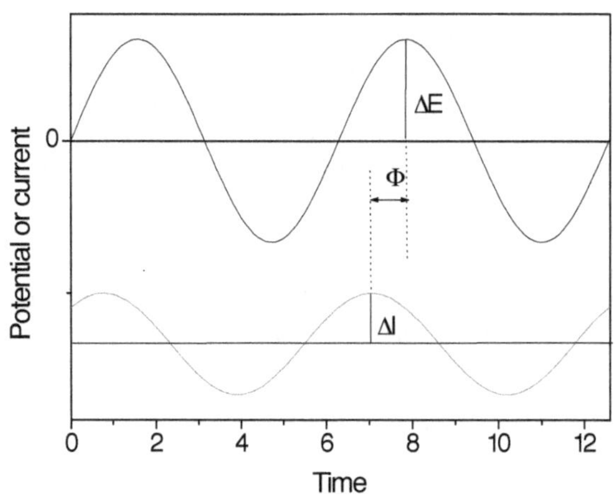

Figure 7-1. Sinusoidal potential and current response signals during electrochemical impedance measurements, where *E* is potential, *I* is current, and ϕ is phase angle.

The applied potential excitation signal is expressed as a function of time by:

$$E(t) = E_0 \sin(\omega t) \tag{7-1}$$

where $E(t)$ is the potential at time t, E_0 is the amplitude of the potential signal, and ω is the radial frequency (radians/s), which is related to frequency, f (Hz) by:

$$\omega = 2\pi f \tag{7-2}$$

The current response signal, $I(t)$, is shifted in phase angle, ϕ, and has a different amplitude l_0:

$$I(t) = l_0 \sin(\omega l + \phi) \tag{7-3}$$

The impedance, Z, is calculated as:

$$Z = \frac{E(t)}{I(t)} = \frac{E_0 \sin(\omega t)}{I_0 \sin(\omega t + \phi)} = Z_0 \frac{\sin(\omega t)}{\sin(\omega t + \phi)} \tag{7-4}$$

Thus, the impedance can be expressed using its magnitude, Z_0, and the phase angle, ϕ.

The impedance can also be expressed as a complex function [Lasia, 1999], where

$$E(t) = E_0 \exp(j\omega t) \tag{7-5}$$

$$I(t) = EI_0 \exp(j\omega t - j\phi) \tag{7-6}$$

$$Z = \frac{E(t)}{I(t)} = Z_0 \exp(j\phi) = Z_0(\cos \phi + j\sin \phi) \tag{7-7}$$

Thus, the electrochemical impedance is composed of a real part, Z_{rea}, and an imaginary part, Z_{ima}, by

$$Z = Z_{rea} + jZ_{ima} = Z_0 \cos\phi + jZ_0 \sin\phi \tag{7-8}$$

$$|Z|^2 = |Z_{rea}|^2 + |Z_{ima}|^2 \tag{7-9}$$

$$\tan\phi = \frac{|Z_{rea}|}{|Z_{ima}|} \tag{7-10}$$

The plot of the real part of impedance against the imaginary part of impedance gives a Nyquist diagram (as shown in Figure 7-2), where the Y-axis is negative and each point on the Nyquist curve is the impedance at one frequency. On the Nyquist plot, the impedance is represented as a vector of length $|Z|$, and the angle between this vector and X-axis (i.e., phase angle ϕ). The advantage of the Nyquist diagram is its quick overview of the impedance data and some qualitative interpretations. When plotting data in the Nyquist diagram, the scale of the real axis must be equal to that of the imaginary axis so as not to distort the shape of the curve, which is important in making qualitative interpretations of the measured data. The disadvantage of the Nyquist diagram is that the data's frequency dimension is not apparently present.

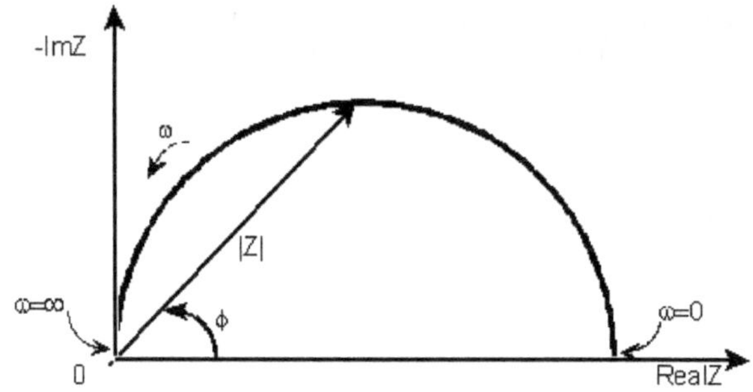

Figure 7-2. A typical Nyquist diagram showing the measured impedance data.

The absolute value of the measured impedance and its phase shift are plotted as a function of the logarithm for measuring frequency in two different plots, giving the Bode plot as shown in Figure 7-3. The Bode plot provides another popular presentation method for electrochemical impedance data.

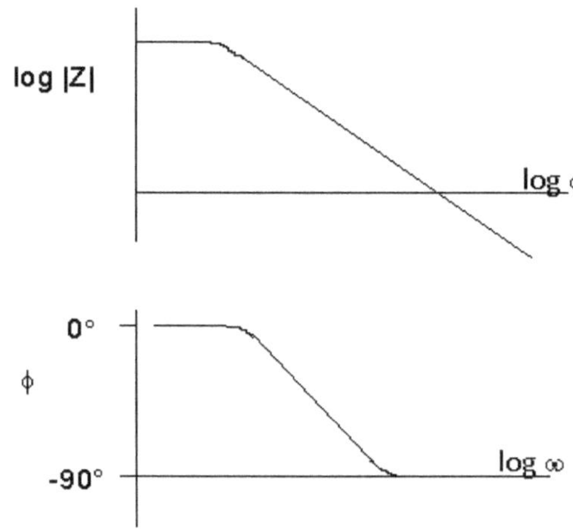

Figure 7-3. A typical Bode plot showing the electrochemical impedance data, where $|Z|$ is the absolute value of impedance, ϕ is phase angle, and ω is radial frequency.

EIS analysis is commonly conducted by fitting the measured impedance data with an electrochemical equivalent circuit, which contains different connections of circuit elements such as resistors, capacitors, and inductors to reflect the physical meaning of the electrochemical system under investigation. The impedances of a resistor with resistance R_e, a capacitor with capacitance C, and an inductor with inductance L_{in} are R_e, $1/j\omega C$, and $j\omega L_{in}$, respectively.

Furthermore, capacitors in EIS measurements often do not behave ideally. Instead, they act like a constant phase element (CPE). These apply to electrodes with inhomogeneous surface conditions, such as steel electrodes containing corrosion pits, coated steel electrodes with coating defects, metal specimens covered with porous corrosion products, etc. The impedance of a CPE, Z_{CPE}, is:

$$Z_{CPE} = \frac{1}{(j\omega)^{\alpha} Y_0} \tag{7-11}$$

where α is an exponent ($-1 \le \alpha \le 1$), denoting the distribution of time constant (α is equal to 1 for a capacitor) and Y_0 is the modulus. The double-layer capacitor in real corrosion cells often behaves like a CPE, *not* a capacitor.

Very few electrochemical systems can be modeled using a single equivalent circuit element. Instead, the circuits usually consist of a number of elements in serial and/or parallel combinations. For linear impedance elements such as Z_1, Z_2, Z_3, ... in series, the equivalent impedance, Z_{eq}, is:

$$Z_{eq} = Z_1 + Z_2 + Z_3 + \ldots \tag{7-12}$$

For linear impedance elements in parallel connection, the equivalent impedance is:

$$\frac{1}{Z_{q}} = \frac{1}{Z_1} + \frac{1}{Z_2} + \frac{1}{Z_3} + \ldots \tag{7-13}$$

For a typical corrosion cell where a metal is actively corroded in electrolyte, the electrochemical equivalent circuit usually includes a parallel connection of a charge-transfer resistance, R_{ct} (sometimes, it is written as polarization resistance, R_p), and a double-layer capacitance, C_{dl}, which is in serial connection with a solution resistance, R_s. The equivalent impedance of the corrosion cell is then written as:

$$\begin{aligned} Z_{eq} &= R_s + \frac{1}{\frac{1}{R_{ct}} + j\omega C_{dl}} = R_s + \frac{R_{ct}}{1 + j\omega R_{ct} C_{dl}} \\ &= \left(R_s + \frac{R_{ct}}{1 + \omega^2 R_{ct}^2 C_{dl}^2} \right) + j \frac{-\omega R_{ct}^2 C_{dl}}{1 + \omega^2 R_{ct}^2 C_{dl}^2} \end{aligned} \tag{7-14}$$

The measured EIS on this corrosion cell can be represented in both the Nyquist diagram and the Bode plot, shown in Figure 7-4. In the Nyquist diagram, the solution resistance is the interception of the real axis with the measured semicircle at the high-frequency end. The real-axis value at the low-frequency intercept is the sum of the charge-transfer resistance and the solution resistance. Thus, the diameter of the semicircle is equal to the charge-transfer resistance. The double-layer capacitance can be obtained from the measuring frequency at the peak of the semicircle, f, by:

$$C_{dl} = \frac{1}{2\pi \hat{f} R_{ct}} \tag{7-15}$$

In Bode plots, the plateau of the absolute value of the measured impedance at the high-frequency end is equal to solution resistance, and that at the low-frequency end is the sum of the solution resistance and charge-transfer resistance, as marked in Figure 7-4c. The maximum number of the phase angle in the phase angle *vs.* frequency plot indicates the number of time constants in the corrosion system. The one maximum of phase angle in Figure 7-4c is associated with the R_{ct}-C_{dl} response occurring at the metal/solution interface.

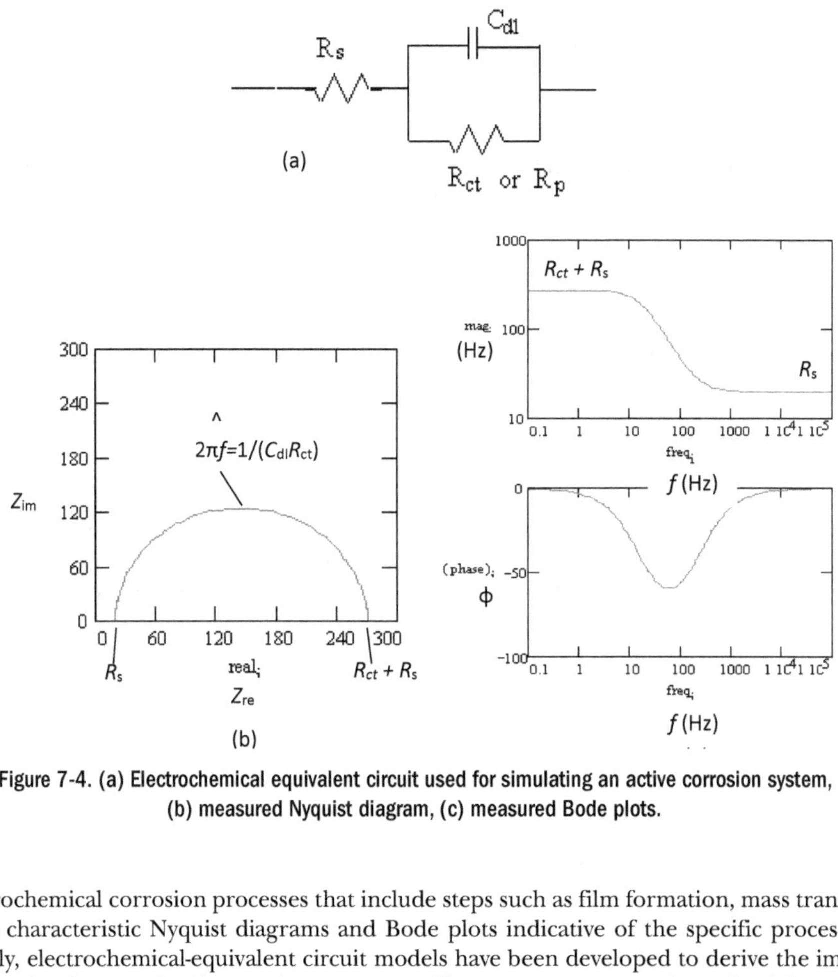

Figure 7-4. (a) Electrochemical equivalent circuit used for simulating an active corrosion system, (b) measured Nyquist diagram, (c) measured Bode plots.

For electrochemical corrosion processes that include steps such as film formation, mass transfer, etc., there are characteristic Nyquist diagrams and Bode plots indicative of the specific process. Correspondingly, electrochemical-equivalent circuit models have been developed to derive the impedance parameters for the associated corrosion processes. There are a great number of articles and books covering this important topic [Mansfeld, 1990; Walter, 1986; Lasia, 1999; Cogger and Evans, 1999]. Therefore, it is not repeated here.

7.1.2. EIS Measurements on Coated Steel Electrodes–Purely Capacitive Coatings

Coatings have been one of the primary techniques used for corrosion control in a wide variety of industrial sectors. As a coating will experience degradation, damage, and failure during service, development of a reliable, rapid method enabling evaluation of the coating properties and prediction of the long-term performance under actual conditions becomes paramount. The EIS technique provides the right solution.

For a metal specimen coated with an intact coating (i.e., no defect is contained in the coating), the measured impedance shows a purely capacitive behavior, with an extremely large impedance value. The equivalent circuit includes a serial connection of a solution resistance and a coating capacitance. The equivalent impedance of the system is equal to $R_s - j\frac{1}{\omega C_c}$, where C_c is the coating capacitance. The ideal Nyquist diagram and Bode plots for this system are shown in Figure 7-5, where the solution resistance is marked. The coating capacitance cannot be determined from the Nyquist diagram, and the low-frequency impedance value is much larger than the solution resistance.

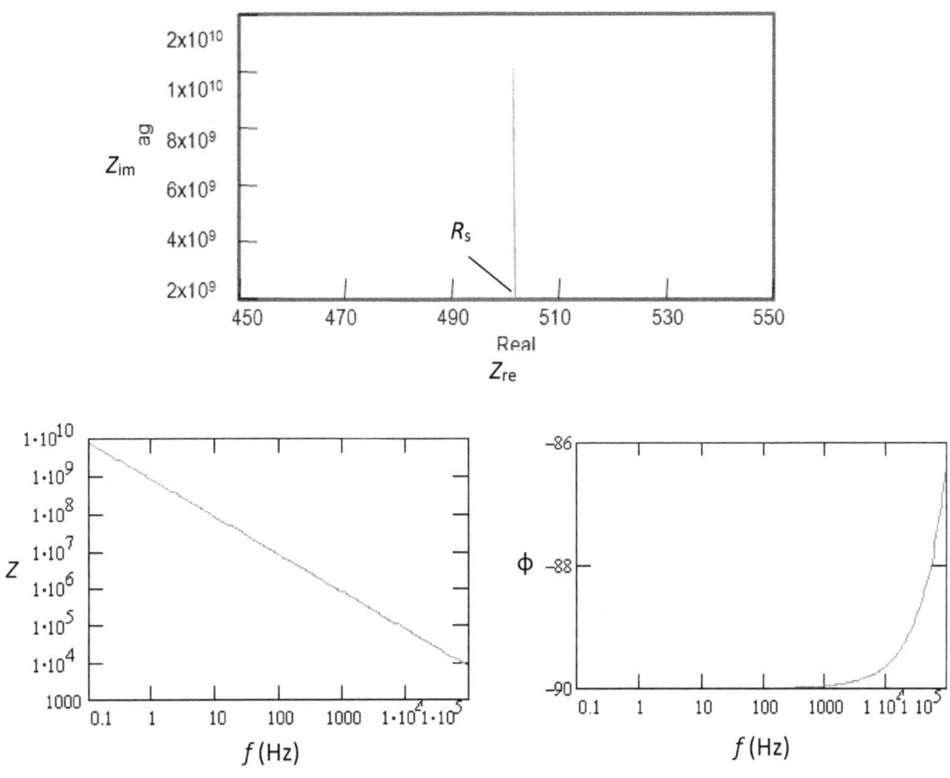

Figure 7-5. Ideal Nyquist diagram and Bode plots obtained on an intact coating.

In reality, the ideal EIS plots in Figure 7-5 are not obtained. When a coated-metal specimen is exposed to aqueous service environments, even if the coating is intact, water may penetrate the coating subject to the coating's water permeability and environmental conditions such as temperature. This would change the shape of the EIS plot. Figures 7-6 and 7-7 show the Nyquist diagram and Bode plots, respectively, measured on an HPCC-coated X65 steel electrode in 0.1 M NaCl solution, where the HPCC coating is intact and does not contain any defect [Howell and Cheng, 2007]. The Nyquist diagram features a big, incomplete semicircle (rather than a vertical line in Figure 7-5), which is caused by the water entry into the coating. Since the coating's impedance is extremely high, the coating is highly resistant and able to protect the steel from corrosion attack. Moreover, the low frequency impedance in the Bode plot becomes flat, rather than a straight line over the whole measuring frequency range in Figure 7-5, indicating that the water penetration tends to trigger reaction at the coating/steel interface.

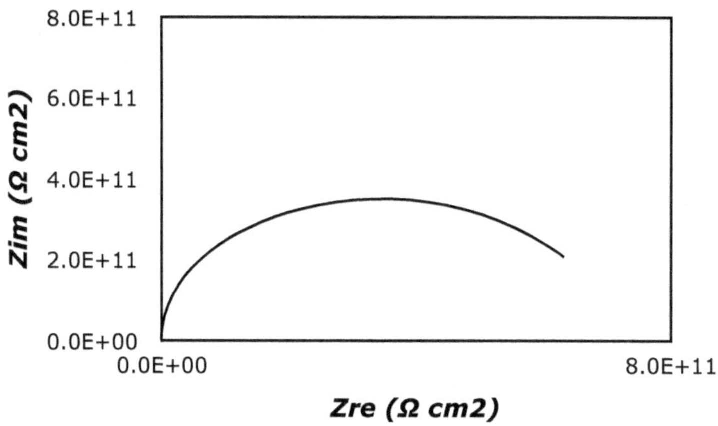

Figure 7-6. Nyquist diagram measured on an HPCC-coated X65 steel electrode in 0.1 M NaCl solution [Howell and Cheng, 2007].

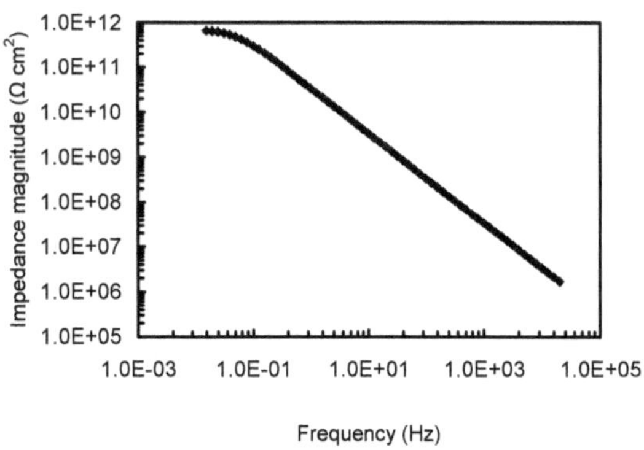

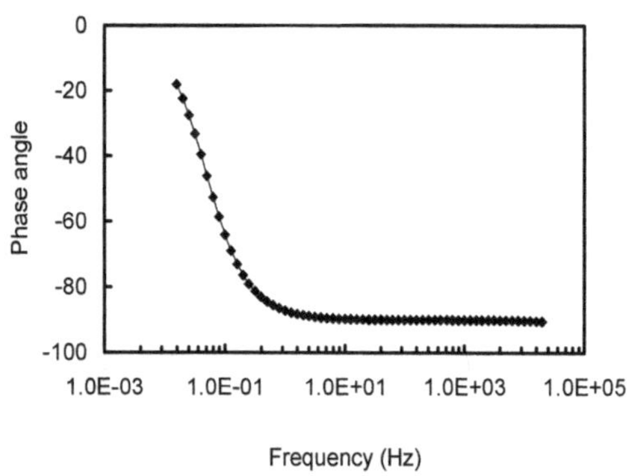

Figure 7-7. Bode plots measured on an HPCC-coated X65 steel electrode in 0.1 M NaCl solution [Howell and Cheng, 2007].

7.1.3. EIS Measurements on Coated Steel Electrodes–Corrosion of Steel beneath Coating

After a certain time period of service in the field, most coatings degrade due to water penetration and, sometimes, due to trapped electrolytes at the coating/metal interface where corrosion occurs. The electrochemical impedance behavior of coated metals (steels) in aqueous environments has been studied extensively. The equivalent circuit shown in Figure 7-8 has been commonly used to fit the measured EIS data. Generally, the coating capacitance is much lower than the double-layer capacitance at the metal/solution interface. The pore resistance, R_{pore}, is the resistance of ion-conducting paths in the coating.

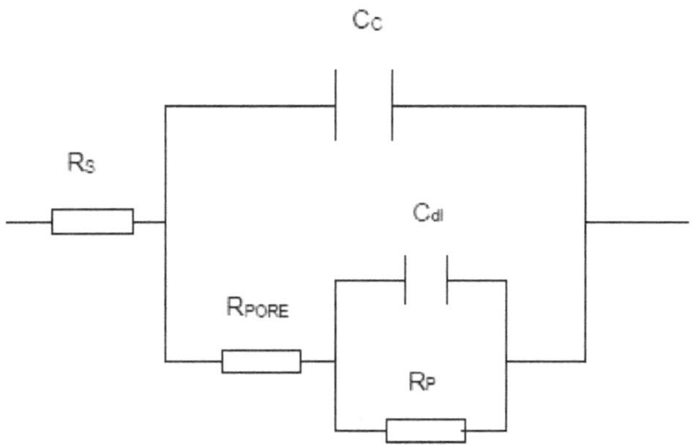

Figure 7-8. Equivalent circuit used to fit the EIS data measured on coated metals, where C_c and R_{pore} are the capacitance and pore resistance of the coating, respectively.

Figures 7-9 and 7-10 show the Nyquist diagram and Bode plots measured on an HPCC-coated steel electrode after 24 h of immersion in a near-neutral pH bicarbonate solution [Zhong et al., 2008]. The Nyquist diagram features two semicircles in the whole measuring frequency range. While the high-frequency semicircle is attributed to the impedance response from the coating, the low-frequency semicircle is associated with the charge-transfer reaction occurring at the steel/coating interface. In the Bode plots, two time constants are identified, as indicated by the two impedance plateaus and two phase-angle maximums. Identical to the Nyquist diagram, the two time constants are attributed to the impedance responses from the coating/solution and steel/coating interfaces, respectively.

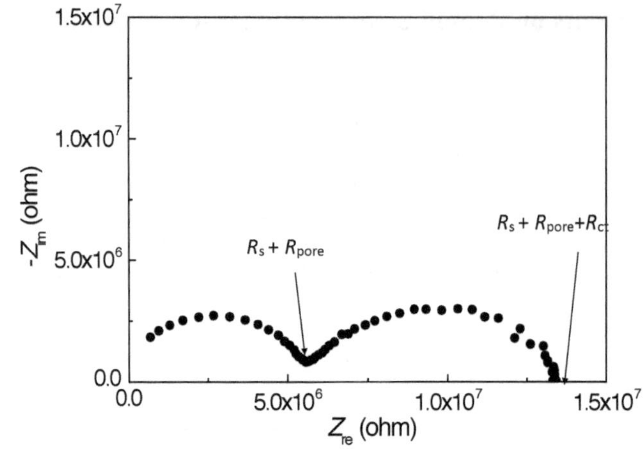

Figure 7-9. Nyquist diagram measured on HPCC-coated-steel electrode after 24 h of immersion in a near-neutral pH bicarbonate solution [Zhong et al., 2008].

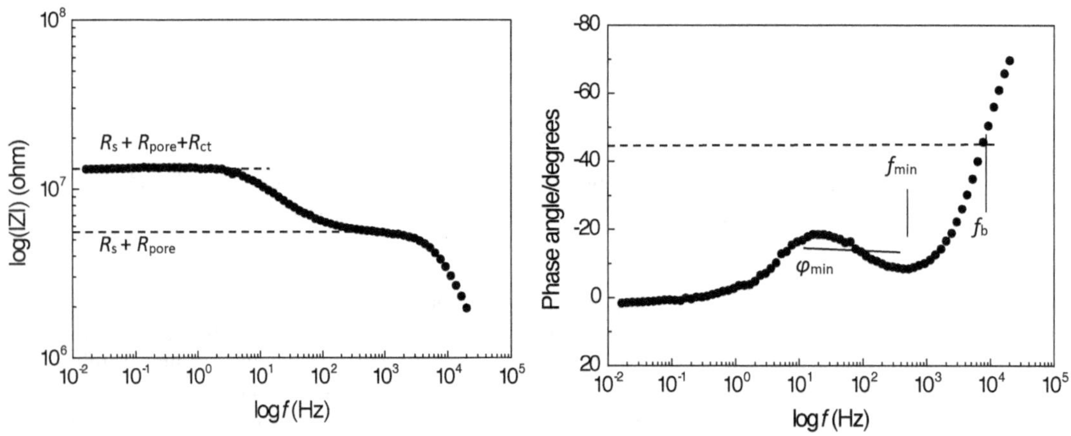

Figure 7-10. Bode plots measured on HPCC-coated steel electrode after 24 h of immersion in a near-neutral pH bicarbonate solution [Zhong et al., 2008].

In addition to performing a mechanistic analysis of the interfacial corrosion reaction (charge-transfer reaction) and the coating properties for coated metals in a corrosive electrolyte, EIS can also derive quantitative information, including charge-transfer resistance, double-layer capacitance, pore resistance of the coating, and coating capacitance by fitting the measured impedance data with the established equivalent circuit. The interception of the high-frequency semicircle with the x-axis at the high-frequency end is the solution resistance. The pore resistance and the charge-transfer resistance are determined in both the Nyquist diagram and Bode plots, as marked in Figures 7-9 and 7-10.

Furthermore, according to Mansfeld's method [Mansfeld, 1995], there are a number of parameters relevant to coating properties; the coating degradation can be derived from the Bode plots.

- Disbonded area (or wetted area) of the coating

The disbonded area, A_d, of a coating can be estimated from R_{pore}, R_{ct}, and C_{dl} by:

$$R_{pore} = \frac{R^0_{pore}}{A_d} \tag{7-16}$$

$$R_{ct} = \frac{R^0_{ct}}{A_d} \tag{7-17}$$

$$C_{dl} = C^0_{dl} A_d \tag{7-18}$$

where R^0_{pore}, R^0_{ct}, *and* C^0_{dl} are the pore resistance, charge-transfer resistance, and double-layer capacitance prior to occurrence of the coating disbondment, respectively.

- Amount of water uptake in the coating

During service of coated structures in aqueous environments, water penetrates the coating and degrades the coating performance. The capacitance of the coating, which can be determined by fitting EIS data with the equivalent circuit in Figure 7-8, is related to the dielectric constant of the coating, ε, by:

$$C_c = \frac{\varepsilon \varepsilon_0 A}{L} \tag{7-19}$$

where ε_0 is the dielectric constant of free space ($\varepsilon_0 = 8.85 \times 10^{-14}$ F/cm), A is the area of the tested electrode, and L is the thickness of the coating film. Generally, the coating capacitance increases with the increasing water uptake. Moreover, it is quite sensitive to the amount of water uptake since the dielectric constant of water is about 20 times larger than that of the coating [Bellucci and Niodemo, 1993]. The volume fraction, v, of water absorbed by the coating can be determined from the experimental values of C_c by [Touhsaent and Leidheiser, 1972]:

$$\upsilon = \frac{\log \frac{C_c(t)}{C_c(0)}}{\log 80} \tag{7-20}$$

where $C_c(t)$ is the coating capacitance at time t and C_c (0) is the initial coating capacitance.

- Breakpoint frequency

The breakpoint frequency, f_b, is the frequency at which the phase angle is equal to 45°, as marked in Figure 7-10. The breakpoint frequency is related to the delaminated area, A_d, or the delamination ratio, D_d, by:

$$f_b = \frac{1}{2}\pi R_{pore} C_c = \left(\frac{1}{2}\pi R^0_{pore} C^0_c\right)\left(\frac{A_d}{A}\right) = \frac{1}{2\pi\varepsilon\varepsilon_0\rho} D_d \tag{7-21}$$

where ρ is the resistivity of the coating.

- Minimum of phase angle and its frequency

In addition to the breakpoint frequency, the minimum of the phase angle (ϕ_{min}) and its frequency (f_{min}), as marked in Figure 7-10, can be used to characterize the coating delamination by [Mansfeld and Tsai, 1991]:

$$f_{\min} = \left(\frac{1}{4}\pi^2 R^2_{pore} C_c C_{dl}\right)^{1/2} = \left(\frac{D_d}{4\pi^2\varepsilon\varepsilon_0 C^0_{dl}\rho^2 d}\right)^{1/2} \tag{7-22}$$

$$\tan\phi_{\min} = (4C_c / C_{dl})^{1/2} = \left(\frac{4\varepsilon\varepsilon_0}{C^0_{dl} d D_d}\right)^{1/2} \tag{7-23}$$

$$\frac{f_b}{f_{\min}} = \left(\frac{C_{dl}}{C_c}\right)^{1/2} \tag{7-24}$$

It was argued [Tsai and Mansfeld, 1993] that, when using Eq. (7-21) to determine A_d, both ε and ρ could change with time. Due to water uptake by the coating, ε will increase and ρ will decrease with the development of conductive paths and defects in the coating. Thus, the f_b is not a constant value. To determine whether an observed change of f_b is due to changes in D_d, ρ, or both, Mansfeld and Tsai [Tsai and Mansfeld, 1993] proposed the use of ϕ_{min} (Eq. 7-23) and the ratio f_b/f_{min} (Eq. 7-24), which are independent of the coating resistivity ρ.

It is also noted that the f_b can only be determined for $\phi_{min} < 45°$. The breakpoint frequency measured for $\phi_{min} > 45°$ does not equal f_b as defined in Eq. 7-21. The discussion about the use of f_b in assessment of coating degradation was conducted, but has not yet reached agreement [Hack and Scully, 1991; Mansfeld, 1992].

The breakpoint frequency method and the extended form make it possible to detect coating damage and the loss of corrosion protection in a short time period. This approach allows the qualitative evaluation of a large number of samples or different areas on a sample in a reasonable time period, and is, therefore, especially useful for corrosion monitoring. However, it does not seem entirely clear what quantity is measured with this method (i.e., whether the area of pores and defects in the coating is determined, or whether the wetted area is where the coating is delaminated and active corrosion occurs). Since the f_b can only be measured for $\phi_{min} < 45°$, for most coatings the lower limit for D_d is about 10^{-4} [Mansfeld, 1995]. This corresponds to a circular area with a radius of about 200 μm for an exposed area of 20 cm^2. Even smaller values of D_d can be determined based on f_{min} and ϕ_{min}. The EIS is thus a very suitable technique for detection of both the degradation of polymeric coatings and the initiation of corrosion at the metal/coating interface at the very early stage.

7.1.4. Case Analysis

The integrity maintenance of North American pipelines depends heavily on a high-performance coating (i.e., FBE). Field measurements were made of barrier properties of the coating on various pipelines owned by Canadian Energy Pipeline Association (CEPA) members after various time periods of service using EIS [King et al., 2002]. It was attempted to determine whether EIS can be used as a tool to assess coating properties in the field, and to investigate the use of EIS as a technique to predict the coating's long-term performance.

Four field trials were performed on FBE-coated pipelines, with Table 7-1 showing characteristics of the various sites investigated and some details of the facilities. Three FBE-coating types were included in the study, with two products produced by manufacturer X (coating X1 and X2), and a third product produced by manufacturer Y.

Table 7-1. Details of the Field Sites Investigated [King et al., 2002].

Site	#1	#2	#3	#4
P/L owner	Company A	Company B	Company C	Company B
Age of line	5 yrs	19 yrs	Pipe coated in 1971, exposed to UV for 22 years before installation in 1993	21 yrs
Product type	X1	X2, plus bituminous rock shield	Y	X1
Coating thickness (mils)	14.9 @ 12 o/c 16.3 @ 3 o/c 15.2 @ 9 o/c	13.2±1.9 (FBE) 100-150 (rock shield)	N.D.	12.4 (mean) 13.8 (max) 10.3 (min)
Pipe size	NPS20	NPS36	NPS30	NPS36
CP conditions (on-potential mV_{CCS})	-	-960 and –1110 (1996 and 1999)	-530 200-300 m from ground bed	-1200 to –1270 (1999-2000), 100 m from rectifier
Site characteristics	Slightly moist clay/sand, hilly, excavation in level area, trench dry	Dry trench, mountainous, rock wall on one side of trench, on upside of hill, good drainage	Mountain valley, site located on 30° slope, muddy at time of excavation	Well drained, at base of 5 m rise
Soil type	Clay/sand	Clay, soil, rock	Wet clay	Clay/sand, stony
Pipe temperature	10-15°C	-	77 km D/S of C/S	-
Depth of cover (m)	~1.1	1.2-1.5 m	~2 m	3-4 m

Figure 7-11 shows the test cells in place on the pipe in the ditch at site #1. The measured EIS at all sites is represented in Bode plots in Figure 7-12, where the impedance response of a fresh X1 FBE coating from a laboratory specimen is also included for comparison. At site #1, an impedance of $>10^{9.7}$ Ω cm² (at 0.1 Hz), which is close to the upper limit of the field-used equipment, is obtained on the pipeline coated with FBE that has been in service for only 5 years. The high impedance indicates that the coating provides an excellent barrier for the coated pipeline. Moreover, the shape of the curve (i.e., a linear dependence of the impedance on frequency with a slope of close to 1) indicates that the coating is acting as a near-perfect capacitor. Although the impedance of the field sample at 0.1 Hz is slightly degraded compared with the laboratory panel, the coating on the pipe still offers excellent barrier properties. At site #2, the FBE-coated line had been in service for 19 years. The impedance at 0.1 Hz was found to exceed $10^{9.5}$ Ω cm² with a purely capacitive behavior. The coating provides excellent barrier properties. At site #3, the Y FBE coating on the pipeline was in poor condition with extensive blistering. EIS was measured on an apparently intact coating location on the pipe surface. The magnitude of the impedance is much lower than those observed from sites #1 and #2, as well as from the laboratory sample. The poorer performance of the field-Y coating is believed to be a consequence of UV degradation of the coating prior to installation, rather than an inferior product formulation. The coating was evaluated to have intermediate-to-poor barrier properties. At site #4, a section of 21-yr-old X1 FBE coating was excavated and found to have extensive blistering all over the pipe surface with a poor adherence of the coating to the pipe surface, believed to be the result of poor application procedures. The low-frequency impedance ($10^{8.4}$ Ω cm² at 0.1 Hz) of the apparently intact coating revealed intermediate-good barrier properties.

Figure 7-11. Soft cells and a calibration cell attached to coated pipe surface at Site #1 for EIS measurements [King et al., 2002].

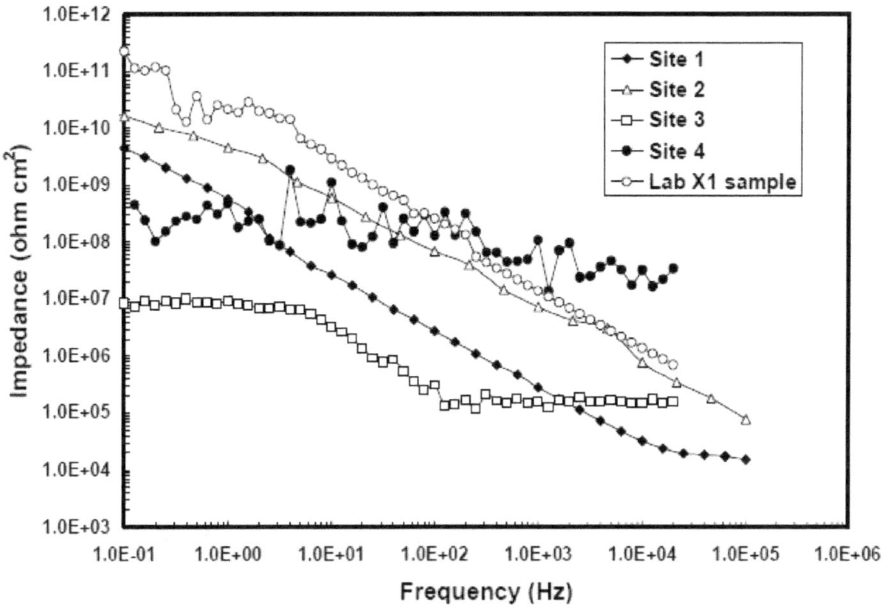

Figure 7-12. Bode plots (the magnitude of impedance) measured on FBE-coated pipelines at various sites and a fresh laboratory sample of FBE coating X1 [King et al., 2002].

Field measurements show that the low-frequency impedance provides a quantitative measure of the coating's barrier properties at the time of inspection, where the coating's general condition can be assessed. Thus, EIS provides a promising method for evaluating coating performance in the field.

Furthermore, analysis of EIS data obtained on coatings with different service years can develop a prediction method for long-term performance of the coating. Figure 7-13 shows the calculated rate of degradation ($-\frac{d\log|Z|}{dt}$) of FBE coatings measured in the laboratory and in various field trials. The data suggest a rapid initial decrease in impedance followed by a slow long-term degradation process. These two degradation modes are believed to be associated with the initial absorption of water (which is dominant during a relatively short term of service) and the long-term degradation of the coating resulting in capillary formation (which is dominant during the longer-term field exposures [King et al., 2002]).

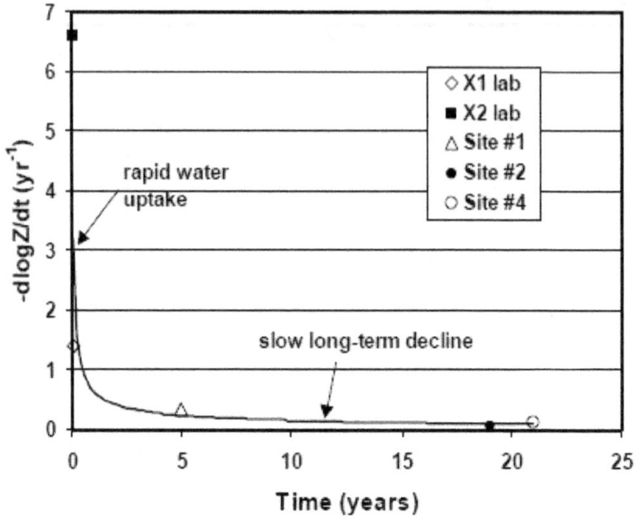

Figure 7-13. Time dependence of the rate of the impedance decrease of FBE coatings tested in both the field trails and in the laboratory [King et al., 2002].

Based on the time dependence of the impedance data, the coating's degradation rate can be derived. The fitted equation from the data in Figure 7-13 is:

$$-\frac{d\log|Z|}{dt} = 0.64t^{-0.641} \tag{7-25}$$

where t is the time with the unit of year. Generally, if the coating degradation is due to the short-term water absorption, the time dependence of $-\frac{d\log|Z|}{dt}$ would be approximately proportional to t^1. The magnitude of the time exponent (-0.61) fitted herein shows that long-term degradation of the coating is occurring. The determination of the exponent -0.61 occurs on the testing results from specific coating films. In other words, the exponent may change when different coating products are measured.

The results and analysis show that, if the impedance of a coating at a given time is known through EIS measurement, the fitted equation similar to Eq. (7-25) can predict the long-term performance of the coating. However, the data shown in Figure 7-13 were obtained from intact coating (i.e., the coating film is intact even when it is disbonded from the substrate, and it does not contain blisters, pinholes, or other defects). Moreover, Eq. (7-25) is useful for estimating the long-term degradation of the coating's barrier properties. As such, it may be useful for predicting future CP requirements as the coating becomes more permeable. At the same time, if the CP current demand increases more rapidly than expected, this may indicate more rapid coating deterioration.

7.2. Localized Electrochemical Impedance Spectroscopy

In the past decades, corrosion research via conventional electrochemical techniques has been bottle-necked by their "spatial resolution" limitation, which disables these techniques from investigating corrosion processes on a microscopic scale, and thus a more mechanistic level. For example, EIS can characterize coating performance and study corrosion of metals beneath the coating [Mansfeld, 1995]. However, the measured impedance results are attributed to electrochemical responses of the whole macroscopic electrode (or specimen) in the environment, thus reflecting an "average" behavior of the whole electrode. When the coating contains micro-defects, such as pinholes, the electrochemical corrosion occurring at the base of the defects is "averaged" out [Zhong et al., 2008]. Obviously, the conventional EIS is incapable of capturing mechanistic information associated with the localized corrosion occurring at the coating defect.

Development of advanced micro-electrochemical techniques, such as LEIS, SKP, etc., has enabled corrosion research with a high spatial resolution. Thus, mechanistic information, which is usually not revealed by conventional macroscopic electrochemical measurements, can advance our understanding of corrosion phenomena, including those occurring beneath disbonded coating and at coating defects.

7.2.1. The Technique and Measuring Principle

Localized Electrochemical Impedance Spectroscopy (LEIS) evolved from the conventional EIS technique, with a similar measuring principle. It enables measurements of the electrochemical impedance (both the impedance modulus and the spectroscopy) specific to local, micron-scaled sites (such as corrosion pits, metallurgical defects, coating defects, etc.). Figure 7-14 shows the schematic diagram of the experimental setup for LEIS measurements on a coated steel specimen. The LEIS measurement system includes three integrated components (i.e., a data acquisition and analysis electrometer, including potentiostat, frequency response analyzer (FRA), and computation unit); a 3-dimenional scanning control station that enables movements of a microprobe at x-, y-, and z-directions; and a specimen-installation component, where a coated steel electrode is used as the working electrode, a SCE as reference electrode, and a platinum wire as counter electrode.

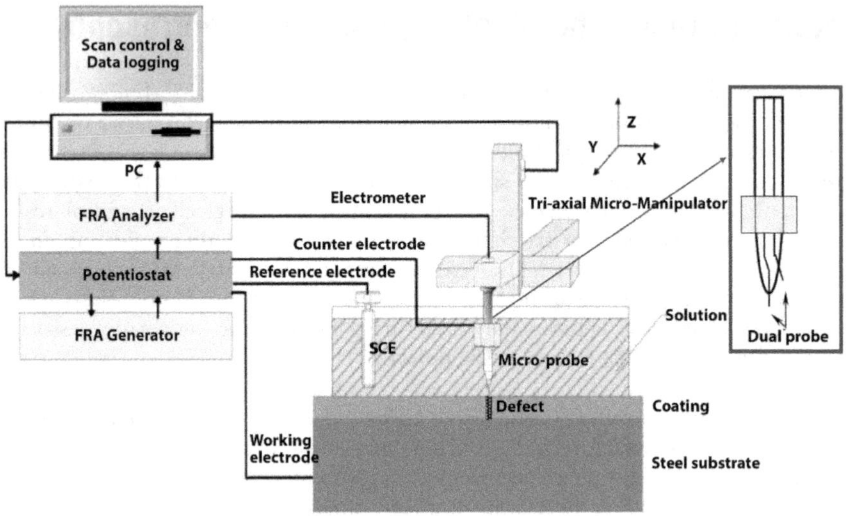

Figure 7-14. Schematic diagram of the experimental setup for LEIS measurements on a coated-steel electrode, where a defect is contained in the coating.

During LEIS measurements, the microprobe, which is made of platinum and is actually a dual probe (as seen in Figure 7-14), is scanned over the surface of the working electrode, measuring the local current density in the electrolyte. The potential difference between the platinum microprobe and another platinum ring electrode contained in the dual probe, ΔE_{local}, is measured via an electrometer. The local AC current density, i_{local}, is calculated, for a known solution conductivity, κ, by

$$i_{local} = \frac{\Delta E_{local}}{d} \times \kappa \qquad (7\text{-}26)$$

where d is the distance between the two platinum electrodes included in the dual probe.

The ratio of the AC voltage perturbation applied on the electrode, E_{appl}, to i_{local} then gives the value of local impedance, Z_{local}, by

$$Z_{local} = \frac{E_{applied}}{i_{local}} \qquad (7\text{-}27)$$

The LEIS microprobe can be operated in two modes. The first mode is used for point-to-point measurements. The microprobe is set directly above the specimen to measure impedance responses at individual points over a certain frequency range. The output is the impedance spectroscopy specific to that point, which can be as small as 10 microns. The second mode is for mapping the amplitude of impedance at a fixed frequency. The microprobe is stepped over a designated area on the specimen. The scanning takes the form of a raster in the x-y plane. The distribution of the impedance magnitude at the fixed frequency within the scanning area is then measured. Thus, this mode delivers the distribution of the impedance modulus at a certain frequency over a certain area or linearly.

7.2.2. LEIS Measurements on Coated Steel Specimens

As discussed, coated-steel pipelines can fail in the field by a number of mechanisms, including localized corrosion of steel at the base of coating defects, missing coating, coating disbonding, corrosion beneath disbonded coating, etc. LEIS is highly suitable for studying the steel corrosion at coating defects, and to evaluate the shielding effect of the defect on CP performance.

The presence of defects with varied sizes is almost an inevitable phenomenon on coatings, including pipeline coatings. Figure 7-15 shows the LEIS maps measured at 50 Hz on an HPCC-coated X65 steel electrode with 200-μm and 1000-μm defects contained in the center of the coating, respectively, in a near-neutral pH bicarbonate solution [Zhong et al., 2008]. The defect is associated with the lowest impedance in the recorded maps. Moreover, based on the color bar that indicates the impedance value, the size of the defect can be estimated. Furthermore, the lowest impedance measured at the defect in Figure 7-15b (about $4.0 \times 10^4\ \Omega$) is smaller than that at the defect in Figure 7-15a (about $1.2 \times 10^5\ \Omega$), indicating that the larger defect in Figure 7-15b has more corrosion activity in the solution than the smaller defect.

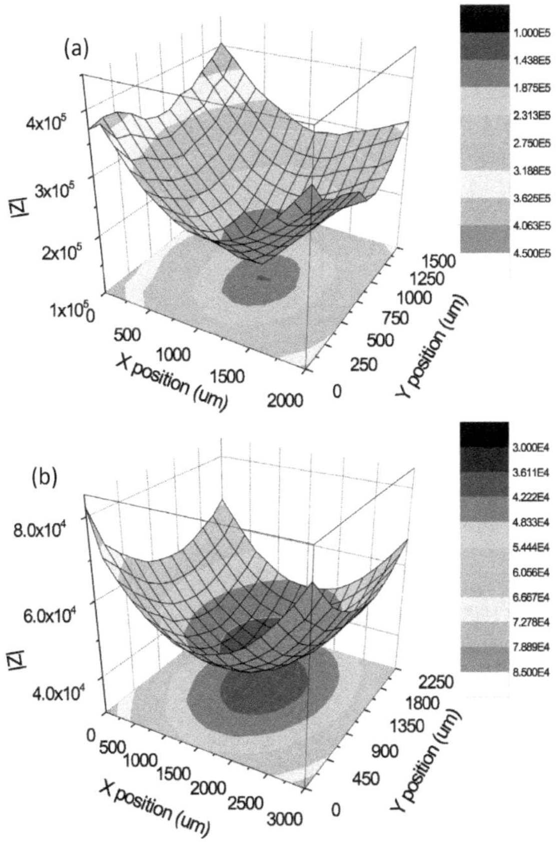

Figure 7-15. LEIS maps measured on an HPCC-coated steel electrode containing a defect of (a) 200 μm and (b) 1000 μm in diameter in a near-neutral pH bicarbonate solution [Zhong et al., 2008].

Figure 7-16 shows the LEIS Nyquist diagrams measured at the defect of 200 μm in diameter after various immersion times in the near-neutral pH bicarbonate solution [Zhong et al., 2008]. It is seen that, after 3 h, the LEIS plot exhibits a capacitive loop over the whole frequency range (Figure 7-16a). With 24 h of immersion, the LEIS plot (Figure 7-16b) is characterized by two depressed semicircles. Upon 48 h of immersion, the impedance plot (Figure 7-16c) contains two semicircles, followed by a straight line with an approximate 45^0 slope in the low-frequency range. When the immersion time increases to 168 h, a linear impedance dominates the low-frequency range and a semicircle exists in the high-frequency range, respectively, as seen in Figure 7-16d.

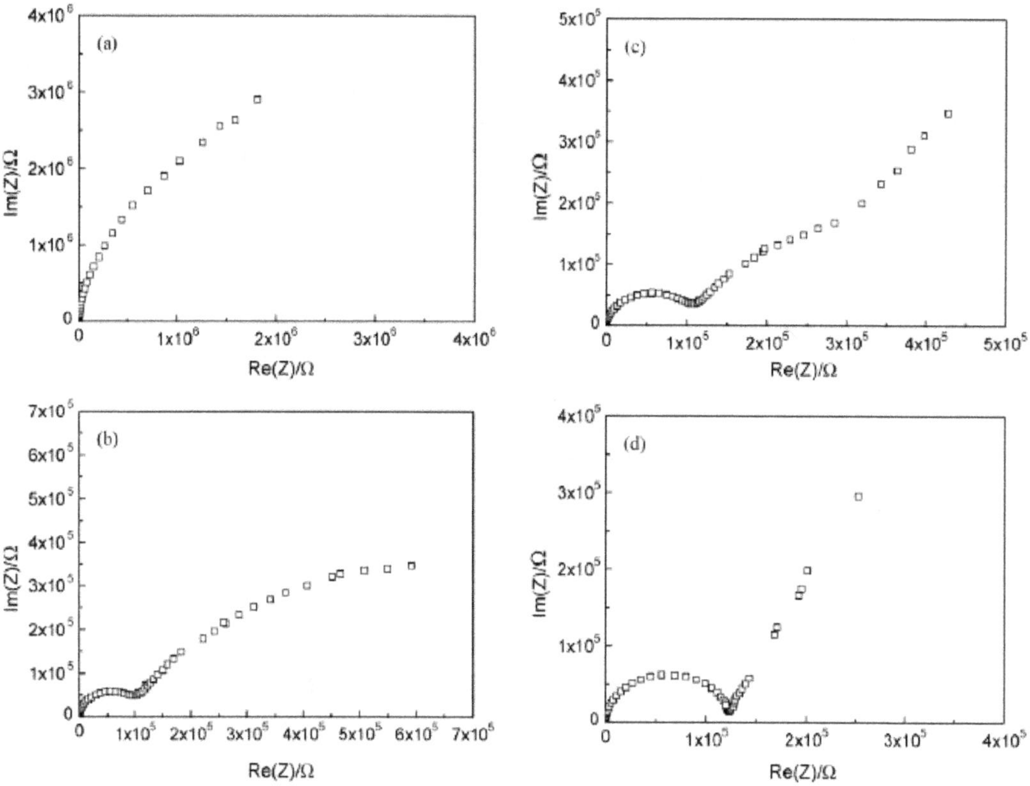

Figure 7-16. Nyquist diagrams of LEIS measured at the defect of 200 μm in diameter after (a) 3 h, (b) 24 h, (c) 48 h, and (d) 168 h of immersion in the near-neutral pH bicarbonate solution [Zhong et al., 2008].

As a comparison, Figure 7-17 shows the Nyquist diagrams measured by a conventional EIS on a coated steel electrode containing a central 200-μm defect at different immersion times (3, 24, 48, and 168 h). It is apparent that the impedance plots measured at individual times are quite different from the LEIS plots shown in Figure 7-16. There is a single, big capacitive loop observed at 3 h of immersion. Upon further immersion, the impedance plots are characterized with two semicircles in the high- and low-frequency ranges. With increasing time, the sizes of both semicircles reduce.

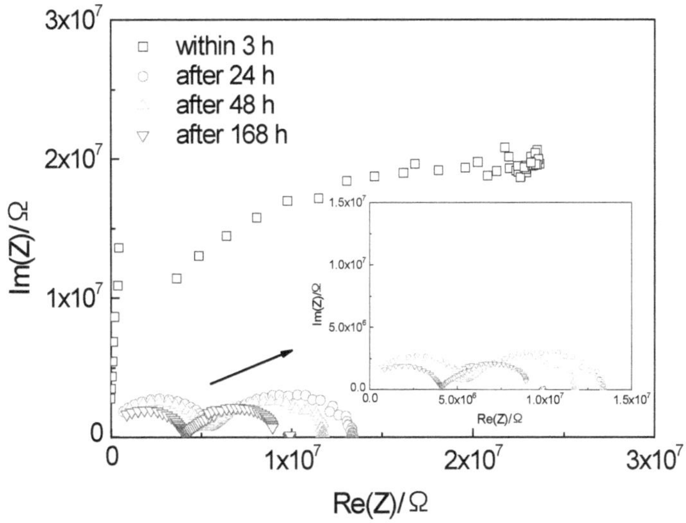

Figure 7-17. Nyquist diagrams measured by conventional EIS on the same electrode as that in Figure 7-16 under the identical testing condition [Zhong et al., 2008].

The Nyquist diagrams obtained by LEIS and conventional EIS at individual times are quite different. The LEIS technique obtains information that cannot be provided by the conventional EIS technique. When a coating contains small pinholes (such as the defect of 200 μm in diameter in this work), the corrosion of steel at the defect experiences mechanistic changes with time, as identified by the distinct LEIS plots measured at individual times. At the early stage of immersion (e.g., after 3 h of immersion of the coated steel electrode in the solution), the LEIS plot shows a capacitive behavior with a large magnitude of impedance. At this stage, due to the small size of the defect, the steel is not yet fully exposed to the electrolyte. It usually takes time for corrosive species to diffuse through the defect to reach its bottom. Therefore, the impedance response is dominated by the dielectric property of the coating.

After 24 h of immersion, two time constants are observed in LEIS plot, as seen in Figure 7-16b. Moreover, the low-frequency impedance decreases dramatically compared to that measured at 3 h of immersion. It is the typical impedance behavior measured on steel beneath a degraded coating [Mansfeld, 1995]. The presence of two time constants indicates that there are two interfacial reaction processes occurring over the measuring frequency range. The high-frequency semicircle is associated with the pore impedance at the defect, which is about 1.5×10^5 Ω, while the low-frequency semicircle is associated with the charge-transfer reaction at the defect base, with a low-frequency impedance value of about 10^6 Ω.

After 48 h of immersion, three frequency-dependent time constants are observed in the Nyquist diagram, as seen in Figure 7-16c. Similarly, the high-frequency semicircle is from the pore impedance at the defect, with an impedance of approximately 1.5×10^5 Ω. The two time constants in the low-frequency range indicate that the corrosion of steel at the defect experiences an activation-diffusion mixed-control process. The straight line with a slope close to 45^0 in the low-frequency range is a Warburg-diffusive impedance. The overlapped Warburg impedance with a semicircle in the low-frequency range shows that the diffusion process contributes to the corrosion reaction at the defect base. Due to the small size of the defect and a relatively long pathway of about 1.1 mm for corrosive species to diffuse from bulk solution to the defect base and, furthermore, due to deposit of corrosion products blocking the diffusive path, the diffusion step becomes dominant over the corrosion process.

With 168 h of immersion, the measured LEIS plot contains two time constants again, with one semicircle at high frequency and a diffusive Warburg straight line at low frequency, as shown in Figure 7-16d. The high-frequency semicircle is from the pore impedance, and the low-frequency Warburg impedance behavior shows that diffusion dominates the corrosion process. The diffusion-controlled effect is primarily due to the blocking effect of corrosion products deposited inside the small defect.

Conventional EIS measurements on the coated-steel electrode that contains a defect provides distinctly different results from those measured by LEIS. In the early stage of immersion, the EIS plot measured after 3 h of immersion is similar to the LEIS plot with the same immersion period (i.e., a capacitive behavior with one time constant over the measuring frequency range). As discussed, the electrolyte has not yet reached the steel at the defect base to result in corrosion. Therefore, the measured impedance is dominated by the coating property at this stage. After 24, 48, and 168 h of immersion, the EIS impedance diagrams feature two semicircles in the high- and low-frequency ranges, respectively, representing the typically reported impedance feature measured on a defect-containing coating systems [Mansfeld, 1993]. The high-frequency semicircle is attributed to the impedance response from the coating, while the low frequency semicircle is associated with a corrosion reaction occurring at the steel/coating interface. Compared with the LEIS plot evolutions with the immersion time, it is obvious that the EIS measurements miss some important information that is directly relevant to the electrochemical corrosion process occurring at the defect. For example, the conventional EIS does not tell the mechanistic aspects regarding the mass-transport process and the blocking effect of corrosion products, as well as the resulting changes of corrosion mechanism. Actually, the conventional EIS measurements give an "averaged" result from both the coating and the defect, and thus, misses the information about the localized corrosion occurring at the coating defect.

7.2.3. LEIS Measurements at Coating Defects

As previously stated, the coating defect, depending on its size and shape, would affect the CP effectiveness locally, and thus, the corrosion of steel at the defect base. To date, a complete understanding of a localized corrosion reaction occurring at the base of the coating defect on cathodically protected pipelines has remained unclear. It is mainly attributed to the limited capability of conventional measurement techniques to characterize localized corrosion processes at a microscopic level [Bayet et al., 1999; Jorcin et al., 2006]. The LEIS technique provides a promising alternative that enables the study of localized corrosion of steel and the CP performance at the base of coating defects.

Figure 7-18 shows the Nyquist diagrams measured by LEIS locally at a defect (200 µm in diameter) on the coated X65 steel electrode at its corrosion potential of -510 mV(SCE) and various cathodic potentials in 0.05 M Na_2CO_3 + 0.1 M $NaHCO_3$ solution. All LEIS plots feature a depressed semicircle in the high-frequency range and an approximately straight line with a 45° slope in the low-frequency range. Furthermore, with the negative increase of the cathodic potentials, the size of the high-frequency semicircle increases. As a comparison, localized EIS plots are measured on bare steel under various cathodic potentials in the same solution, and the results are shown in Figure 7-19. With the increase of cathodic potential, the charge-transfer resistance decreases. The quite different impedance behaviors measured at the coating defect and on the bare steel under the identical cathodic polarization condition shows that the defect would shield the permeation of the applied CP from reaching the steel at the defect base, resulting in different electrochemical responses (as shown in Figures 7-18 and 7-19).

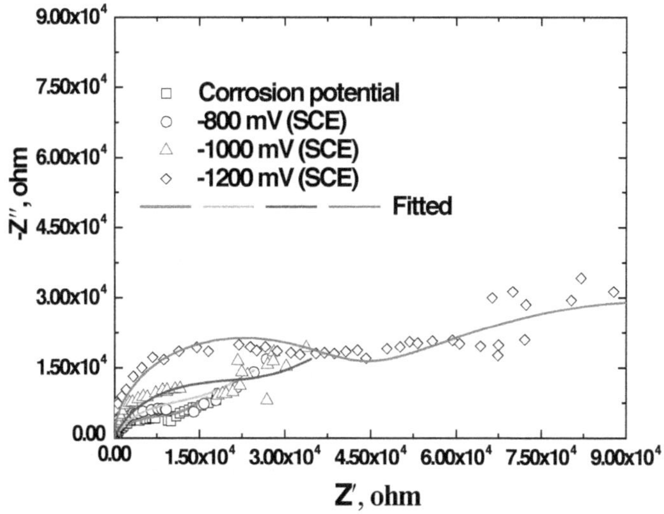

Figure 7-18. Nyquist diagrams measured by LEIS locally at a defect (200 μm in diameter) on the coated X65 steel electrode at its corrosion potential of -510 mV(SCE) and various cathodic potentials in 0.05 M Na_2CO_3 + 0.1 M $NaHCO_3$ solution [Dong et al., 2008].

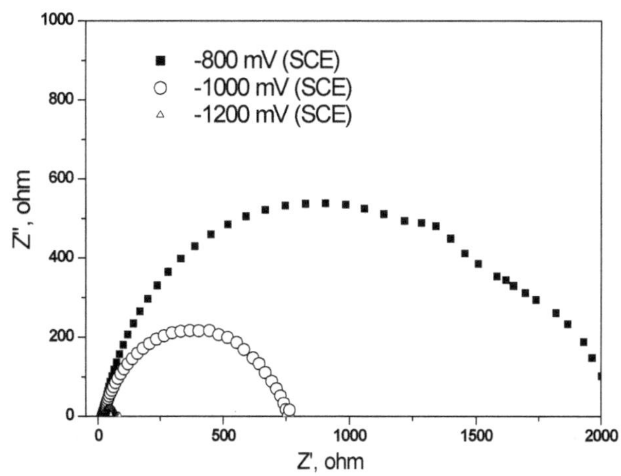

Figure 7-19. Nyquist diagrams measured by LEIS on a bare steel electrode under various cathodic potentials in 0.05 M Na_2CO_3 + 0.1 M $NaHCO_3$ solution [Dong et al., 2008].

For the impedance measured on the bare steel, at -800 mV (SCE), the charge-transfer resistance is from a complex combination of most cathodic reactions (e.g., cathodic reductions of oxygen and/or water) with some anodic reactions (e.g., oxidation of iron). When the cathodic potential is −1000 mV(SCE), the charge-transfer resistance mainly reflects the cathodic reaction, resulting in a significant decrease of the resistance. With the further negative increase of cathodic potential to -1200 mV(SCE), the cathodic-reduction reaction is further enhanced. Thus, there is a further decrease of the charge-transfer resistance, as indicated by the reducing size of the semicircle with the applied cathodic potential.

For LEIS measured at the coating defect, the localized EIS plots show, at OCP and various cathodic potentials, that the localized impedance plots are featured with a high-frequency semicircle. This semicircle is attributed to the interfacial charge-transfer reaction occurring at the base of the defect and a low-frequency straight-line with a 45° slope, which indicates that the mass-transfer step dominates the corrosion process. In general, the mass-transfer step involves diffusion of reactants, such as dissolved oxygen, through the solution layer inside the defect that has a narrow, deep geometry (the depth/width ratio of defect is about 5.5 in this work). The cathodic polarization curve measured on a macroscopic steel electrode with a defect in the coating shows clearly the presence of a cathodic-limiting diffusive current, as shown in Figure 7-20. This is consistent with the measured diffusive impedance behavior. Furthermore, even at a very negative potential (e.g., -1200 mV(SCE)), the measured impedance spectroscopy is still associated with the diffusion-controlled corrosion process at the defect base.

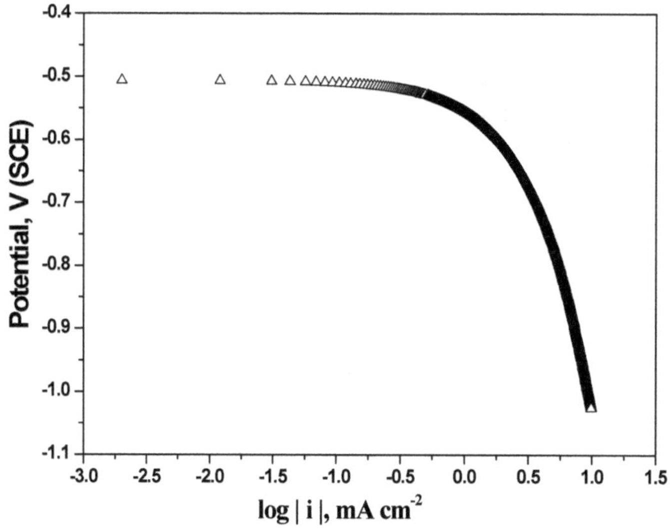

Figure 7-20. Cathodic polarization curve measured on the coated steel electrode containing a coating defect under the identical condition to that in Figure 7-18 [Dong et al., 2008].

Generally, the impedance predominantly reflects the property of the partial reaction that carries the majority of current. At CP potentials, especially at the over-protected potential such as −1200 mV(SCE), it is the cathodic reaction that carries the majority (or even all) current for a bare-steel electrode. In comparison to the electrochemical impedance plots measured at the coating defect and on the bare steel in Figures 4-18 and 4-19, the applied CP cannot fully penetrate the defect to reach the steel at the defect base. The CP current is shielded, at least partially, due to the narrow, deep geometry of the defect. Therefore, CP shielding occurs not only under disbonded coating (as discussed in Chapter Four), but also at coating defects that are featured with a narrow, deep geometry.

7.3. Scanning Kelvin Probe

7.3.1. The Technique and Measuring Principle

The Scanning Kelvin Probe (SKP) is a non-contact and non-destructive technique that enables mapping and measurements of the difference in work functions between a material in testing and a reference probe. The major advantage of SKP compared to other electrochemical devices is that the Kelvin probe measures electrode potential without touching the surface under investigation across a dielectric medium of a high resistance. Figure 7-21 shows the schematic diagram of the setup for SKP measurements, where the Kelvin probe, which is installed in a three-dimensional scanning control station, is operated in the non-contact mode above a steel electrode in the electrolyte trapped beneath disbonded coating.

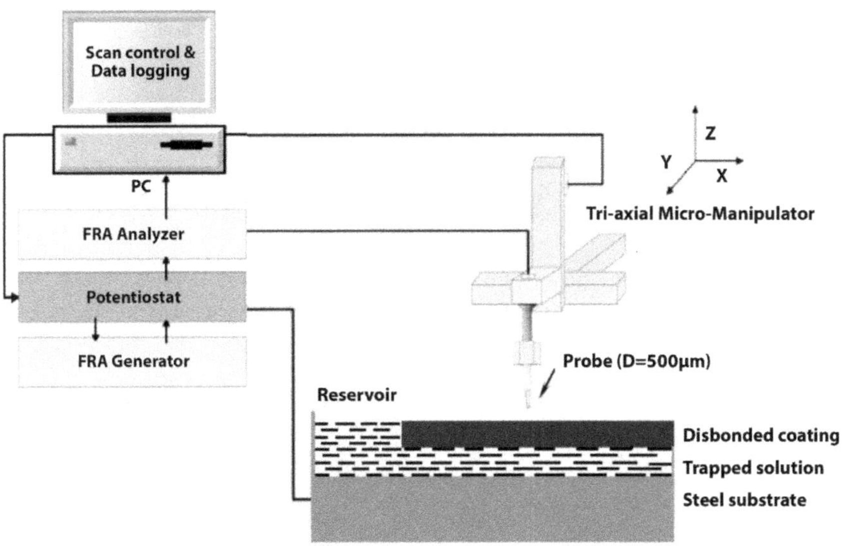

Figure 7-21. Schematic diagram of the setup for SKP measurements via the non-contact mode above a steel electrode in the electrolyte trapped beneath disbonded coating.

When a target electrode is immersed in a certain aqueous environment, there is a work function difference between the electrode and the reference electrode (i.e., the Kelvin probe). Figure 7-22 shows schematics of the SKP working principle. During SKP measurements, the two electrodes are electrically connected by an external electric circuit, where electrons flow in the circuit to establish an equilibrium of charge. The work function, ϕ_W, of a solid material at a solid/liquid interface can be divided into two components [Leng et al., 1999]:

$$\phi_W = \Delta\phi + V_c \quad (7\text{-}28)$$

where $\Delta\phi$ is contact potential, and V_c is surface potential. Since electrons move between the testing electrode and the reference probe through several interfaces, such as metal/solution interface, solution/coating interface, etc., the work function is extremely sensitive to surface conditioning, and is affected by surface phenomena such as adsorption or absorption, surface charging, oxidization, surface/bulk contamination, and corrosion.

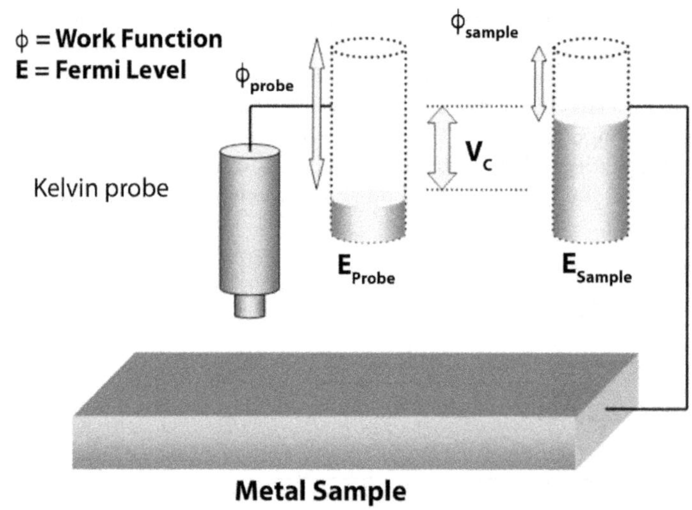

Figure 7-22. Schematic diagram of the working principle of SKP measurements on a metal sample.

Under certain circumstances, the work function is determined by electrode potential. Therefore, the Kelvin probe can measure local corrosion potential of the target electrode in the environment by [Furbeth and Stratmann, 2001]:

$$E_{corr} = \text{constant} + \Delta\phi_W \quad (7\text{-}29)$$

where $\Delta\phi_W$ is the measured work-function difference between the Kelvin probe and the target material.

When SKP measurements are conducted on a coated-steel specimen, the recorded Kelvin potential, $\Delta\psi^{probe}_{coating}$, contains multiple components across various materials and interfaces, as shown in Figure 7-23. Based on situations when corrosion occurs beneath the coating or when corrosion products deposit on the steel, the Kelvin potential is associated with different definitions.

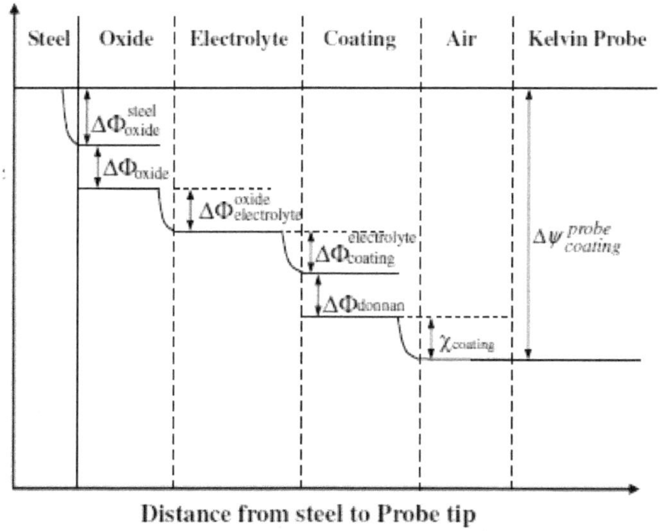

Figure 7-23. Schematic diagram of the multiple components contained in Kelvin potential measured on a coated-steel specimen.

For an intact coating on steel, the measured Kelvin potential is defined as:

$$\Delta\psi^{probe}_{coating} = \Delta\Phi^{steel}_{coating} + \chi_{coating} - \frac{1}{F}(\phi^{probe}_{W} + \mu^{steel}_{e}) \tag{7-30}$$

If the coating is disbonded from the steel, but corrosion products are not yet deposited, the Kelvin potential is:

$$\Delta\psi^{probe}_{coating} = \Delta\Phi^{steel}_{electrolyte} + \Delta\Phi^{electrolyte}_{coating} + \chi_{coating} + \Delta\Phi_{donnan} - \frac{1}{F}(\phi^{probe}_{W} + \mu^{steel}_{e}) \tag{7-31}$$

When corrosion occurs beneath a disbonded coating and corrosion products (such as iron oxide) are formed, the Kelvin potential is:

$$\Delta\psi^{probe}_{coating} = \Delta\Phi^{steel}_{oxide} + \Delta\Phi^{oxide}_{electrode} + \Delta\Phi_{oxide} + \Delta\Phi^{electrolyte}_{coating} + \chi_{coating} + \Delta\Phi_{donnan} - \frac{1}{F}(\phi^{probe}_{W} + \mu^{steel}_{e}) \tag{7-32}$$

where $\Delta\phi^i_j$ is the contact potential at the individual interface between *i* and *j* materials, $\chi_{coating}$ is the surface potential of the coating, $\Delta\phi_{donnan}$ is the Donnan potential that is generated due to the presence of two solution phases separated by the disbonded coating film, *F* is Faraday's constant, ϕ_{probe} is the work function of the reference Kelvin probe, and μ_e^{steel} is the chemical potential of the steel electrode. Thus, the corrosion potential of the steel in aqueous environments can be obtained by measuring the Kelvin potential and the difference of work functions between the Kelvin probe and the steel by combining Eq. (7-29) with Eqs. (7-30) – (7-32).

7.3.2. Monitoring of Coating Disbondment by SKP

Uses of SKP technique in the study of coated metal corrosion focus primarily on three aspects (i.e., monitoring of coating disbondment, characterization of corrosive environments beneath disbonded coating, and measurements of electrochemical corrosion of metals in a thin layer of electrolyte trapped beneath the coating).

The SKP is effective for investigating coating delamination. It was found [Leng et al., 1999b] that there is a clear potential transition line separating the intact area from the disbonded area on the coated metal specimen. This potential line is indicative of the delaminated borderline of the coating. A shift of the borderline into the intact area indicates the extension of coating disbondment occurring at the steel/coating interface. The coating delaminating rate can be calculated according to the displacement of borderline over time.

Figure 7-24 shows schematics of an artificial assembly containing both intact and disbonded regions on a coated-steel electrode [Fu and Cheng, 2009]. The surface marked as "Solution" simulates a disbonded FBE coating, with the disbonding thickness of about 60 μm, on an X65 pipeline steel. The solution contained 0.05 M Na_2CO_3 and 0.1 M $NaHCO_3$, with a pH of 9.6. The red square marked as "Scanning area" refers to the area the SKP has scanned, which includes both intact and disbonded areas.

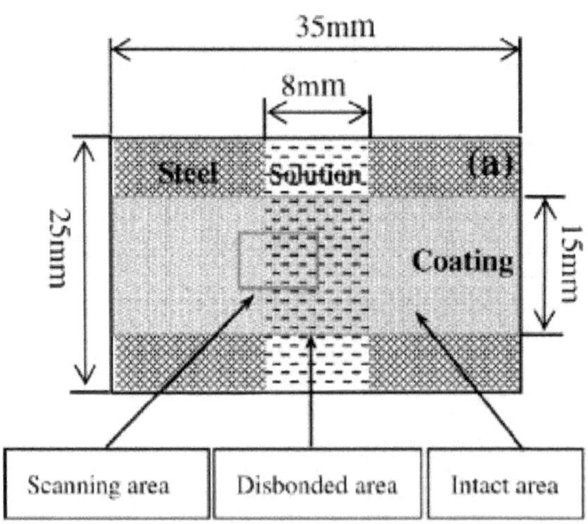

Figure 7-24. Schematic diagram of an artificial assembly containing both intact and disbonded regions on a coated-steel electrode [Fu and Cheng, 2009].

Figure 7-25 shows Kelvin potentials measured on the scanned area (marked in Figure 7-24) as a function of time, where blue and green potential regions refer to disbonded and intact areas, respectively. All potential maps possess a similar feature (i.e., a gradual shift of the Kelvin potential from the intact to the disbonded areas). Moreover, the Kelvin potential of the intact area is always less negative than that of the disbonded area. With the increasing time, potentials measured on both sides shift negatively. For example, the Kelvin potential of the intact area near the disbonding boundary line is about -750 mV (tungsten, W) at day 1, then shifts to a more negative value of -1150 mV (W) at day 9. Similarly, the Kelvin potential of the disbonded area decreases from -1000 mV (W) at day 1 to -1400 mV (W) at day 9. After the SKP measurement, the coated-steel specimen is viewed by an optical microscope (see Figure 7-26). The regions I, II, and III refer to open area, disbonded area, and intact area, respectively. The intact area is penetrated by the aqueous solution, and there are apparently corrosion products present in region II (i.e., the disbonded area).

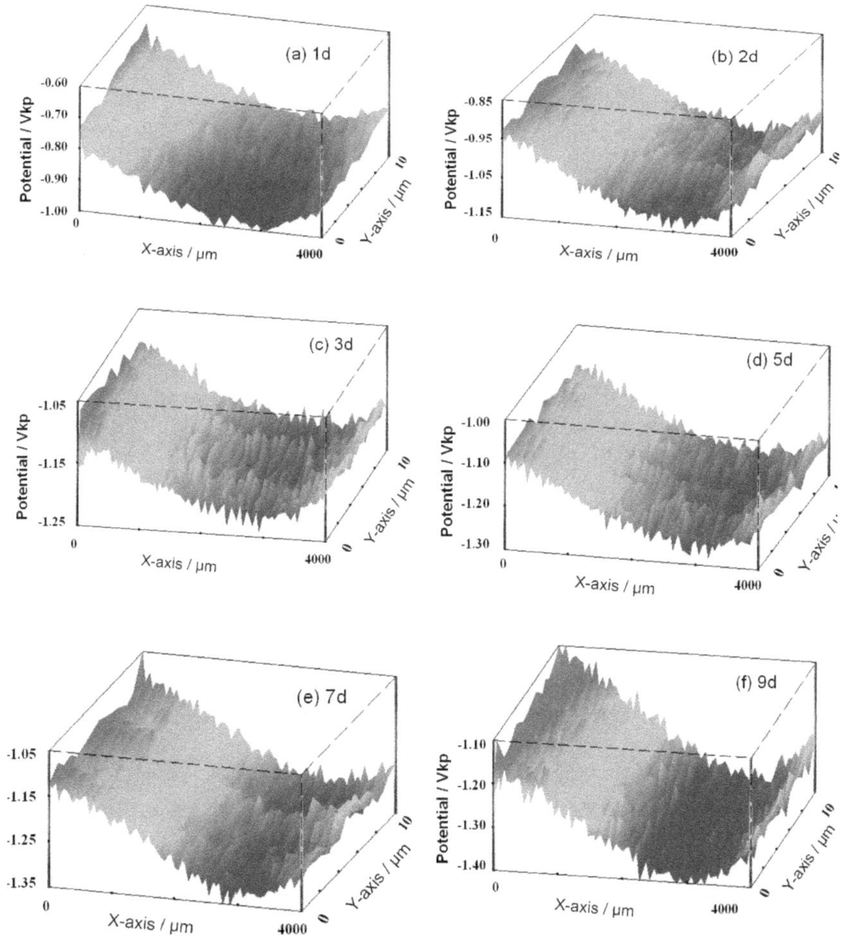

Figure 7-25. Kelvin potentials measured on the scanned area of the specimen in Figure 7-24 as a function of time, where blue and green potential regions refer to disbonded and intact areas, respectively [Fu and Cheng, 2009].

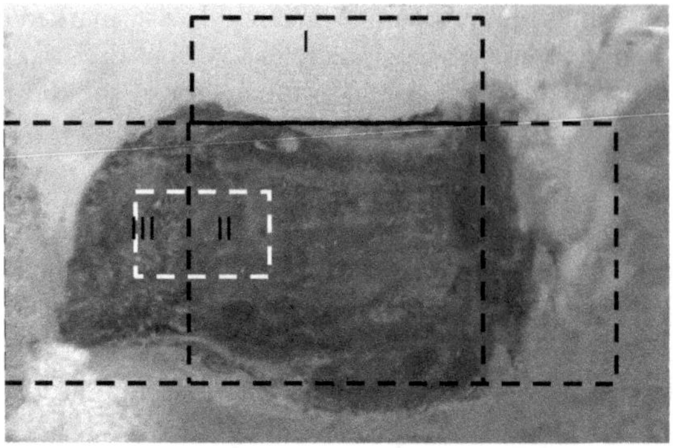

Figure 7-26. Optical view of the coated-steel specimen after SKP measurements [Fu and Cheng, 2009].

In reality, a distinct boundary line of the Kelvin potential for distinguishing intact and disbonded areas on a coated specimen (like those included in Leng et al.'s work [Leng et al., 1999b]) is difficult, if not impossible, to measure. Figure 7-25 shows the results of a gradual increase in the Kelvin potential from the disbonded area to intact area. There is not a distinct, steep boundary line separating the two areas observed. This is primarily due to there being no true "intact" area on the coated-metal specimen, and with the electrolyte penetrating the coating gradually. Analysis of the Kelvin potential identifies the environmental evolution with time beneath the coating. The results in Figure 7-25 show that the Kelvin potential measured on the intact area shifts negatively with time, which is probably due to changes in the electrolyte concentration and/or the electrochemical reaction rate beneath the coating. The Kelvin potential of the intact area decreases remarkably in the first 3 days, and is maintained with a relatively steady value during the following days, which indicates that the continuous solution intake occurs mainly within 3 days. Afterwards, the trapped solution reaches a saturation status. With an increasing amount of the solution, it is expected that both the electrolyte concentration and the electrochemical reaction rate change, resulting in a significant decrease of the interfacial potential. Moreover, both the surface potential of the coating and the Donnan potential decrease with the increase of the water uptake. With continued immersion, the double-charge layer achieves a relatively steady state, and so does the steel/electrolyte interfacial potential. Thus, the change of Kelvin potentials during the following days is slight. There is a more negative Kelvin potential on a disbonded area than there is on an intact area, which is attributed to corrosion occurring beneath the disbonded coating, just like what is shown in Figure 7-26.

7.3.3. Characterization of Corrosive Environments beneath Disbonded Coating by SKP

The SKP is highly sensitive to changes in the surface condition of metals, including corrosion of the metal. When corrosion occurs beneath disbonded coating, corrosive environments may change with time due to a number of factors (such as temperature fluctuations, changes of solution chemistry, aeration and deoxygenation, etc.). For example, Furbeth and Stratmann [Furbeth and Stratmann, 2001] investigated the influence of oxygen on coating disbondment by measuring the change of Kelvin potential of steel beneath disbonded coating. Results demonstrated that the partial pressure of oxygen affects the potential distribution and cathodic reaction kinetics beneath a coating, and thus the disbonding rate of the coating.

To investigate the effect of dissolved oxygen and its diffusion on steel corrosion beneath disbonded coating, a coated electrode assembly (shown in Figure 7-27) was built: A 60-µm thick disbondment of FBE coating was created on pipe steel to make the solution (i.e., a near-neutral pH bicarbonate solution and dissolved oxygen penetrated the disbonding crevice). The SKP measurements were conducted on the red-marked area.

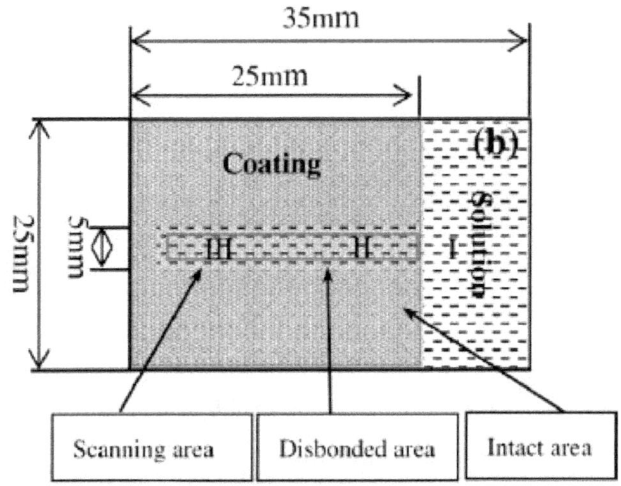

Figure 7-27. Schematic diagram of an artificial assembly simulating FBE-coating disbondment from pipe steel, where the solution and dissolved oxygen penetrate the disbonding crevice beneath the coating. The red-colored area is the SKP scanning and measurement area [Fu and Cheng, 2009].

Figure 7-28 shows the SKP measurement results of the testing assembly as described in Figure 7-27. When farther away from the bulk solution (region I), the Kelvin potential shifts positively. For example, in region III, that is about 12,000 µm from the bulk solution, with the Kelvin potential approximately -0.3 V (W). Closer to the bulk solution, the Kelvin potential decreases. In region II that is about 9,000 µm from the bulk solution, with the average Kelvin potential about -0.6 V (W).

When high-purity nitrogen purges the solution of oxygen, SKP measurements are performed on the same setup. The measured Kelvin potential as a function of time is shown in Figure 7-29. Within 2.5 h of deoxygenation, the potential difference between regions II (closer to the bulk solution) and III (farther from the bulk solution) is about 0.2 V. After 2.5 h, the potential difference decreases to less than 0.1 V. After 5 h, there is an identical Kelvin potential throughout the disbonded area. With increasing time, the Kelvin potential shifts positively. Figure 7-30 shows the linear profile of the Kelvin potential along the disbonded depth as a function of time.

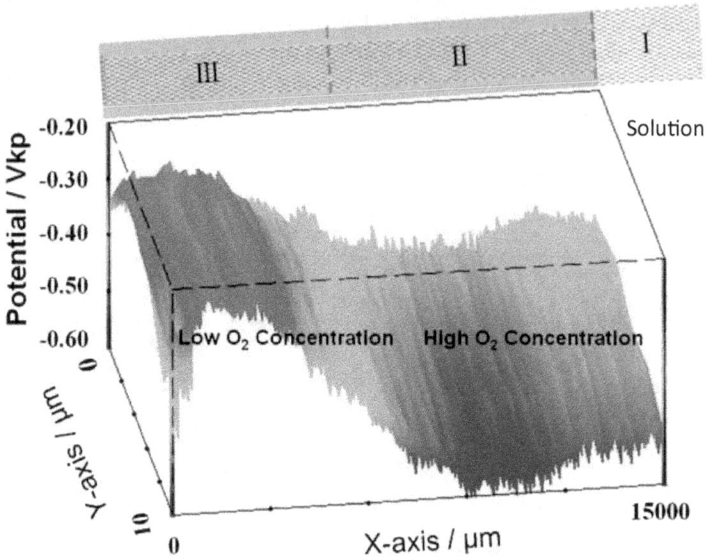

Figure 7-28. Kelvin potential map measured on the specimen shown in Figure 7-27 [Fu and Cheng, 2009].

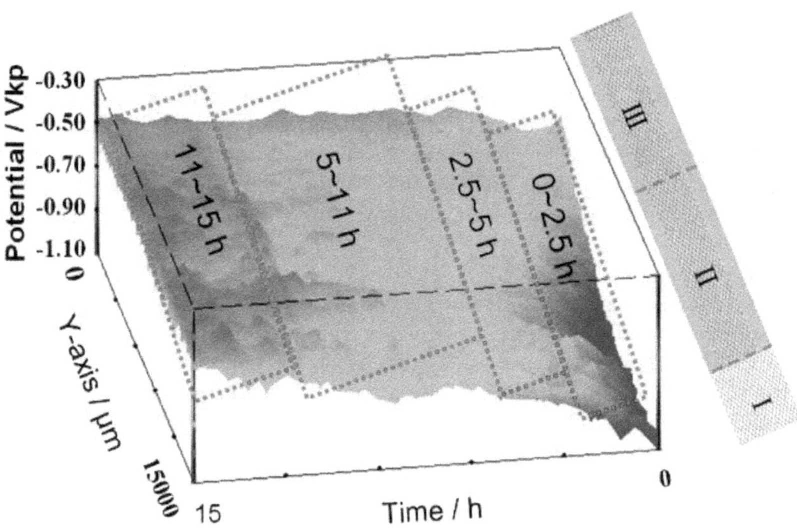

Figure 7-29. Kelvin potentials measured on the same setup in
Figure 7-27 that is purged with nitrogen as a function of time [Fu and Cheng, 2009].

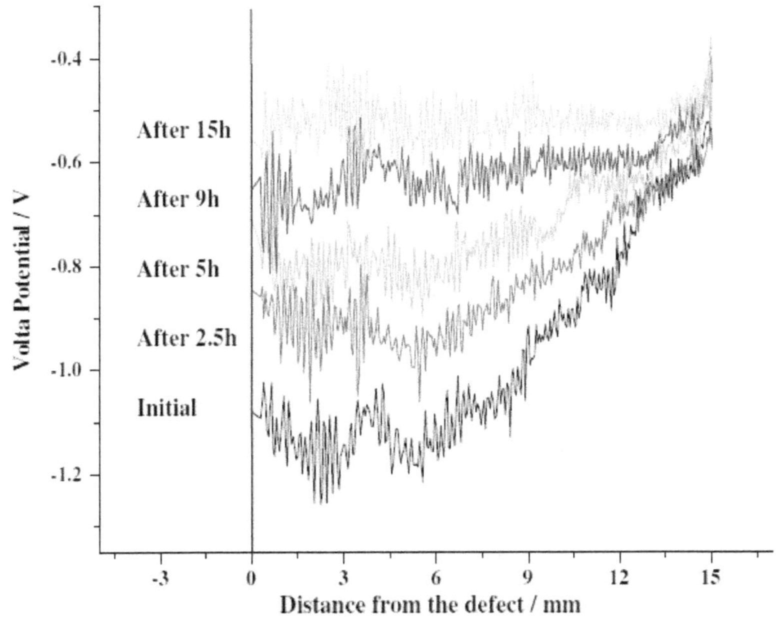

Figure 7-30. Linear presentation of the Kelvin potential measured in Figure 7-29 [Fu and Cheng, 2009].

Due to the narrow geometry of the coating disbondment, an oxygen-concentration difference exists along the direction of the disbonding crevice depth. According to the oxygen-concentration cell theory, area I in Figure 7-27 is open to air and is thus associated with a high oxygen concentration, where the cathodic reduction of dissolved oxygen occurs. Areas II and III behave as an anode where the anodic dissolution of iron occurs because of the low oxygen concentration. Similarly, there is a higher oxygen concentration at area II than at area III. Thus, in a comparison of these two areas, area II tends to be the location for cathodic reaction to occur, and area III is where the anodic reaction occurs. As shown in the Kelvin potential distribution, the potential of area II is more negative than that of area III. In the absence of environmental oxygen, the effect due to the oxygen concentration difference among areas does not exist, and the potential tends towards a positive value, as shown in Figure 7-30. Therefore, the corrosion of steel beneath a disbonded coating strongly depends on the environmental oxygen concentration. When a trace amount of oxygen initially exists in the solution, a potential difference can be recorded between areas II and III. The cathodic and anodic reactions tend to be separated between these areas beneath the disbonded coating. With the nitrogen's continuous removal of oxygen, the Kelvin potential becomes identical throughout the disbonded area. Thus, corrosion, once occurring, is uniform beneath the coating disbondment.

Furthermore, seasonal wet-dry cycles are critical to the development of corrosive environments beneath disbonded coating. During the wet–dry cycling, the thickness of the solution layer trapped beneath disbonded coating decreases due to water evaporation. The reduction of the solution thickness facilitates oxygen diffusion, and simultaneously, causes the solution concentration to increase, resulting in a reduction of the oxygen solubility. While the former contributes to oxidation of the steel and thus an increase in electrode potential, the latter is always related to a negative shift of potential due to less oxygen getting involved in the corrosion of the steel. There usually exists a competition between the two effects, resulting from the seasonal wet-dry cycling on corrosion of the steel. The SKP enables a direct detection of the environmental change during wet-dry cycles, and the resulting effects on corrosion of the substrate steel beneath the coating. Figure 7-31 shows the experimental setup for the investigation of coating disbondment during wet-dry cycles, where the red-marked area is for SKP measurements. The trapped electrolyte contains 0.05 M Na_2CO_3 and 0.1 M $NaHCO_3$, with a pH of 9.6.

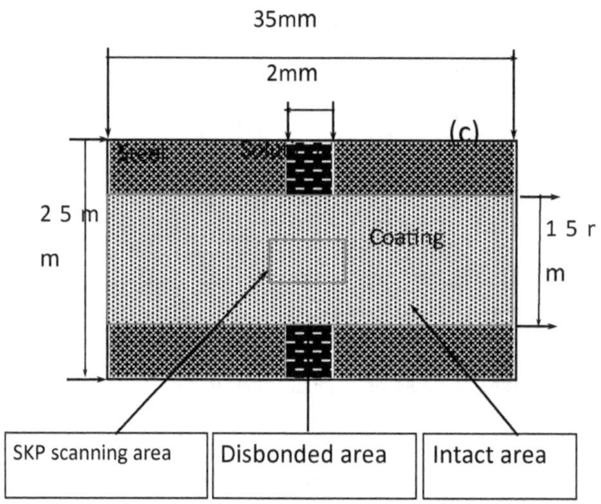

Figure 7-31. Schematic diagram of an artificial assembly used for investigation of coating disbondment during wet-dry cycles [Fu and Cheng, 2009].

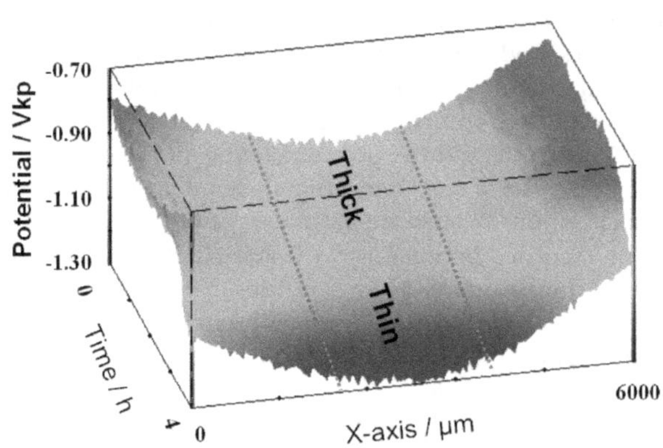

Figure 7-32. Kelvin potential measured on coated steel specimen shown in Figure 7-31 during wet-dry cycling [Fu and Cheng, 2009].

Figure 7-32 shows the SKP potential measured on the coated X65 steel specimen (open to air) experiencing wet–dry cycling, where the low potential region in the central part of the scanning area is the disbonded area. With evaporation of the trapped electrolyte over time, the solution layer changes from "thick" to relatively "thin." Correspondingly, the Kelvin potential shifts negatively. For example, a Kelvin potential of about -1.1 V (W) is observed over a certain area of disbondment when the solution layer is "thick." Over time, the solution evaporates due to reduced environmental humidity. The Kelvin potential drops as the solution layer gradually thins. In the "thin" area, the Kelvin potential is maintained at about -1.3 V(W). A thin-solution layer favors diffusing the dissolved oxygen towards the steel surface for a cathodic reaction. Obviously, the SKP can characterize the corrosion activity of the steel in response to the thickness of the trapped solution layer beneath a disbonded coating during seasonal wet-dry cycles.

References

Baker Jr., M. (2004) *Stress Corrosion Cracking Study*, Integrity Management Program Delivery Order DTRS56-02-D-70036, Department of Transportation, USA.

Bayet, E., Huet, F., Keddam, M. (1999) Local electrochemical impedance measurement: scanning vibrating electrode technique in ac mode, *Electrochim. Acta* 44, 4117-4127.

Bellucci, F., Nicodemo, L. (1993) Water transport in organic coatings, *Corros.* 49, 235-247.

Cheng, Y.F. (2010) Pipeline Engineering, in: *Pipeline Engineering*, Ed. Yufeng F. Cheng, in: Encyclopedia of Life Support System (EOLSS), Developed under the Auspices of the UNESCO, EOLSS Publishers, Oxford, UK.

Cheng, Y.F. (2011) Application of micro-electrochemical techniques in corrosion research, in: *Green Corrosion Chemistry and Engineering*, S.K. Sharma, Editor, Wiley-VCH Publisher, Germany, p. 71-96.

Cheng, Y.F. (2013) *Stress Corrosion Cracking of Pipelines*, John Wiley Publishing, Hoboken, NJ, USA.

Cheng, Y.F., Niu, L. (2007) Mechanism for hydrogen evolution reaction on pipeline steel in near-neutral pH solution, *Electrochem. Commun.* 9, 558-562.

Cogger, N.D., Evans, N.J. (1999) *An Introduction to Electrochemical Impedance Measurement*, Technical Report No. 6, Solartron Analytical, Farnborough, UK.

Frankel, G.S. (2008) Electrochemical techniques in corrosion: Status, limitations, and needs, *J. ASTM Inter.* 5, 1-27.

Furbeth, W., Stratmann, M. (2001) The delamination of polymeric coatings from electrogalvanised steel – a mechanistic approach. Part 1. Delamination from a defect with intact zinc layer, *Corros. Sci.* 43, 207-227.

Hack, H.P., Scully, J.R. (1991) Defect area determination of organic coated steels in seawater using the breakpoint frequency method, *J. Electrochem. Soc.* 138, 33-40.

Howell, G.R., Cheng, Y.F. (2007) Characterization of high performance composite coating for the northern pipeline application, *Prog. Organ. Coat.* 60, 148-152.

Jorcin, J.B., Aragon, E., Merlatti, C., Pébère, N. (2006) Delamination areas beneath organic coating: a local electrochemical impedance approach, *Corros. Sci.* 48, 1779-1790.

King, F., Cheng, Y.F., Gray, L., Drader, B., Sutherby, R. (2002) Field assessment of FBE-coated pipelines and the implications for stress corrosion cracking, *4th Inter. Pipeline Conf.*, IPC02-27100, ASME, Calgary, Canada.

Lasia, A. (1999) Electrochemical impedance spectroscopy and its applications, in: *Modern Aspects of Electrochemistry*, B. E. Conway, J. Bockris, R.E. White, Eds., Kluwer Academic/Plenum Publishers, New York, USA, Vol. 32, p. 143-248.

Leng, A., Streckel, H., Stratmann, M. (1999a) The delamination of polymeric coatings from steel. Part 1. Calibration of the Kelvin probe and basic delamination mechanism, *Corros. Sci.* 41, 547-578.

Leng, A., Streckel, H., Stratmann, M. (1999b) The delamination of polymeric coatings from steel. Part 2: First stage of delamination, effect of type and concentration of cations on delamination, chemical analysis of the interface, *Corros. Sci.* 41, 579-597.

Mansfeld, F. (1990) Electrochemical impedance spectroscopy (EIS) as a new tool for investigation methods of corrosion protection, *Electrochim. Acta* 35, 1533-1544.

Mansfeld, F. (1992) Comment on "Defect Area Determination of Organic-Coated Steels in Seawater Using the Breakpoint Frequency Method", *J. Electrochem. Soc.* 139, 639-640.

Mansfeld, F. (1993) Models for the impedance behavior of protective coatings and cases of localized corrosion, *Electrochim. Acta* 38, 1891-1897.

Mansfeld, F. (1995) Use of electrochemical impedance spectroscopy for the study of corrosion protection by polymer coatings, *J. Appl. Electrochem.* 39, 187–202.

Mansfeld, F., Tsai, C.H. (1991) Determination of coating deterioration with EIS: I. Basic relationships, *Corros.* 47, 958-963.

Stansbury, E.E., Buchanan, R.A. (2000) *Fundamentals of Electrochemical Corrosion*, ASM International, Materials Park, OH, USA.

Touhsaent, R.E., Leidheiser Jr., H. (1972) Capacitance–resistance study of polybutadiene coatings on steel, *Corros.* 28, 435-440.

Tsai, C.H., Mansfeld, F. (1993) Determination of coating deterioration with EIS: Part II. Development of a method for field testing of protective coatings, *Corros.* 49, 726-737.

Walter, G.W. (1986) A review of impedance plot methods used for corrosion performance analysis of painted metals, *Corros. Sci.* 26, 681-703.

Zhong, C., Tang, X., Cheng, Y.F. (2008) Corrosion of steel under the defected coating studied by localized electrochemical impedance spectroscopy, *Electrochim. Acta* 53, 4740-4747.

8 Coating Application on Pipelines

After selection of a coating for a specific environment and before application, a set of concise and thorough specifications must be written. When a series of candidate coating systems emerges, laboratory screening tests should be conducted to rank the candidate systems for potential suitability under the field conditions anticipated in the pipeline service. [Tator, 2006] For the best coating performance, the application process must have a thorough and rigorous but fair inspection by qualified inspectors. Well-trained inspectors with a passion for ensuring a high-quality product are critical in the coating selection process, whether in the field or the plant. [Norsworthy, 2007]

8.1. Specifications

A well-written set of coating specifications is critical to ensuring a coated structure has the expected life and service. A coating specification is the guide for the applicator to know the coating process specifics required by the owner. There are many NACE International coating standards that can be used for basic guidelines for each type of coating system to be used. Be sure to use the most recent revision of the NACE standards.

A process to pre-qualify an applicator is a critical specification step that is many times overlooked, especially with field-applied coating systems. Use pre-qualification to verify that the coating crew in the field or in the plant are capable of applying the chosen coating system. The coating-crew foreman and members must be present during the qualification and throughout the job once the crew passes the pre-qualification for the selected coating type. If a foreman or crew member quits during the project, the new crew must requalify to apply the coating.

The specification is also a reference for any problems, test type and frequency, test evaluation, rejection/acceptance criterion, etc. The specification should include coating thickness parameters, the type of test equipment used, repair materials, etc. Shipping, handling, and storage should also be a significant part of the coating specification. Inspection parameters should be spelled out in the specification. Inspectors (especially the lead inspector) should have experience with the particular type of coating system being used by the applicator, as well as be NACE-certified.

Specifications should be clear and concise so the inspector knows what is expected and can work with an applicator to ensure the best possible coating system for the owner. Specifications should give the inspector all guidance possible for making fair decisions about any testing and application processes.

The above information is limited. Each specification should be written for a specific project and coating system chosen.

8.2. Surface Preparation Overview

The success of practically every coating system depends upon the initial cleaning and preparation of the surface to be coated. Surface preparation is a coating system's foundation, and is critical to coatings used with CP.

Surfaces must be properly prepared before protective coatings are applied for good results to be obtained. Most pipeline coatings require both a clean surface and good anchor pattern for long-term durability and adhesion. Adhesion is important for coated, buried, or submerged structures that are cathodically protected. Most premature coating failures are caused either completely or in part by inadequate or improper surface preparation. Loss of adhesion can lead to CP shielding, as discussed earlier.

The activities of surface preparation prior to coating application include:

- Assessment or inspection of surface conditions, including design and fabrication defects
- Pre-cleaning or removal of visible and invisible surface deposits such as oil, grease, and various salts
- Work to remedy or alleviate design or fabrication defects such as pipe-surface slivers and gouges
- Inspection and documentation of the pre-cleaning process and cleaning defects, if any
- Surface preparation by appropriate method to remove detrimental surface contaminants

Many factors in surface preparation affect the life of a coating, including:

- Residues of oil, grease, and soil, which can prevent adhesion or mechanical bonding of the coating to the surface
- Residues of chemical salts (such as chlorides or sulfates), which can induce corrosion if water penetrates or moves under the coating. These may be invisible, but several test methods are available for various salts.
- Rust on the surface, which interferes with how coating bonds to the surface
- Loose or broken mill scale can cause early coating failure because the mill scale is not adhered.
- Tight mill scale is also a problem, since mill scale is cathodic to the pipe steel and corrosion will develop if water penetrates
- Anchor patterns (formed by surface preparation actions), which may be so rough that peaks are formed that are difficult to adequately protect with coatings or does not allow the coating to properly fill the valleys

- If the profile is not deep enough, then loss of adhesion may occur.
- Sharp ridges, burrs, edges, or cuts from mechanical-cleaning equipment that prevents consistent coating thickness
- Surface condensation, which may result in blistering and delamination
- Old coatings that may have poor adhesion or may be too deteriorated for recoating
- Existing coatings that may be incompatible with or affected by the application of coatings

Surfaces to be coated often require pre-cleaning. Inspection for contamination, including deposits of grease, oil, dust, dirt, or salts, is an important part of the overall coating process. Contaminants may be visible or invisible.

Inspection for surface cleanliness is a continuous process, and should take place at least three times during the coating process (on each pipe or field joint):

- Before any surface preparation activities
- After surface preparation, before coating begins
- Between each application of coating in a multi-coat system

8.2.1. Surface Cleanliness

Surfaces must be free of oil and grease before blast cleaning. Blast cleaning will not remove oil and grease, but will spread grease and oil over the surface and contaminate abrasive-blast media, which is important if abrasives are recycled.

Abrasive blasting will not remove surface salts. Salts may be visible or invisible contaminants and are best removed by water washing, high-pressure water blast, or acid washing. All surfaces should be inspected after cleaning for specification compliance.

8.2.2. Surface Preparation Standards and Procedures

The most common surface preparation standards have been prepared by NACE International, SSPC, and ISO. These standards are compared in Table 8.1.

Table 8.1. Comparative Listing of NACE, SSPC, and ISO Surface-Preparation Standards (Note: This chart is comparative only, since many standards are not equivalent).

	NACE	SSPC	ISO 8501-1
	NONABRASIVE CLEANING		
Solvent Cleaning		SSPC-SP-1	
Hand Tool Cleaning		SSPC-SP-2	St2 or St3
Power Tool Cleaning		SSPC-SP-3	St2 or St3
Power Tool Cleaning to White Metal		SSPC-SP-11	
Flame Cleaning		SSPC-SP-4	F1
Pickling		SSPC-SP-8	
Water Jetting	NACE No. 5/SSPC SP-12		
	ABRASIVE BLAST CLEANING		
	JOINT SURFACE PREPARATION STANDARDS		
White Metal	NACE No. 1/SSPC-SP-5		Sa 3 ("Blast-Cleaning to Visually Clean Steel")
Near-White Metal	NACE No. 2/SSPC-SP-10		Sa 2 ½ ("Very Thorough Blast-Cleaning")
Commercial	NACE No. 3/SSPC-SP-6		Sa 2 ("Thorough Blast-Cleaning")
Brush-Off	NACE No. 4/SSPC-SP-7		Sa 1 ("Light Blast-Cleaning")
Industrial	NACE No. 8/SSPC-SP-14		

8.2.3. Blast Cleaning

Though it is called "blast cleaning," the term does not always mean the surface is clean. Most coatings perform better on grit-blasted surfaces than surfaces prepared by power or hand tools. [Norsworthy; D'Ambrosio; Quinn; 2015] As mentioned above, invisible contaminants can be spread through blasting. These must be removed by proper washing or solvent cleaning before blasting. Some blast-cleaning methods include:

- Dry-abrasive blast
- Centrifugal blast
- Abrasiveinjected water blast
- Slurry blast
- Wet-abrasive blast

8.2.3.1. Dry Grit Blast Cleaning

The most generally established method of surface preparation for coating application is dry-grit blast cleaning. Dry-grit blasting uses a highly concentrated stream of small abrasive particles projected, usually by compressed air, at a surface to remove rust, mill-scale, or other contaminants while creating a rough surface that is beneficial for coating adhesion. The profile created by grit blasting exposes more surface area for the coating adhesion, compared to an un-blasted surface. For most pipeline coatings, there is no truly satisfactory or economically equivalent alternative to grit blasting.

The fundamental process of the grit-blasting process is the removal of rust, mill-scale, or other surface contaminants while obtaining a suitably roughened surface by projecting a highly concentrated stream of relatively small abrasive particles at high velocity against the surface to be cleaned. The surface is abraded through the high-velocity impact of abrasive particles. To prepare steel surfaces for coating, blast cleaning removes rust, mill-scale, and old paint along with a very small amount of the base metal.

Various degrees, or standards, of surface cleanliness achieved by abrasive blast have been defined. The most commonly used abrasive-blast cleaning standards for new steel are produced by NACE, SSPC, and ISO. NACE and SSPC issued a number of joint surface preparation standards for abrasive-blast cleaning, including:

- NACE No. 1/SSPCSP 5 "White Metal Blast Cleaning," reaffirmed in 1999
- NACE No. 2/SSPCSP 10 "NearWhite Metal Blast Cleaning," reaffirmed in 1999
- NACE No. 3/SSPCSP 6 "Commercial Blast Cleaning," reaffirmed in 1999
- NACE No. 4/SSPCSP 7 "BrushOff Blast Cleaning," revised in 2000
- NACE No. 8/SSPC-SP 14 "Industrial Blast Cleaning," new in 1998

These standards are roughly equivalent to the ISO standards that were developed from the original Swedish standards. ISO 8501-1 was published in 1988 and contains four standards:

- Sa3 "Blast-Cleaning to Visually Clean Steel"
- Sa2½ "Very Thorough Blast-Cleaning"
- Sa2 "Thorough Blast-Cleaning"
- Sa1 "Light Blast-Cleaning"

Each standards system represents a progressive scale of visual appearance of only the best grade being shown first. The quality of blast cleaning is determined visually, with photographic standards generally used for comparison purposes. There is no correlation between the degree of blast cleaning used and the surface profile produced; there is no specific correlation with the removal of chemical contamination (or invisible salts). For these issues, other standards and measuring techniques must be used.

8.2.3.2. Blast-cleaning Equipment

Abrasive-blast media are projected by the direct feed of particles from a pressurized container into a high-pressure air stream (pressure blasting) or by the centrifugal projection from rapidly rotating impellers (centrifugal blasting or airless blasting). Pressuring blasting is the most commonly used method of abrasive blasting in the field. Abrasive-blast media are forced under pressure from the pressure vessel (blast pot) through the blast hose. This is a high-production method for field and shop applications and can be used on various irregular-shaped objects such as valves.

Blast-cleaning equipment typically consists of an air compressor, blasting pot, and blasting hose. A vacuum unit enables material that is blasted forward to be recovered immediately. This method is used when fugitive airborne abrasive particles are undesirable (for example, near sensitive equipment). This blasting method is expensive, slow, and is used only in special situations. It is rarely used on pipelines.

Blast-cleaning cabinets are sometimes desirable for blast-cleaning individual items such as induction bends, valves, and other components in an enclosed space. Companies that blast and coat various pipeline components will buy or build a blast-cleaning cabinet. Cabinet size may vary from the very small cabinet, where blasting is done from outside the cabinet with hands inserted through handholes in the side, to the relatively large blast room. The more sophisticated blast rooms may use a rail system to transport large items into the blasting area and have grit recovery and recycling systems. In general, the blast-cleaning apparatus is similar to that used for on-site blasting.

The most complex blast-cleaning cabinets are designed for large quantities of steel to be blast-cleaned on a regular basis, such as all pipe in a pipe-coating mill. These machines are designed to work on continuously and include a conveyor system that continuously carries items through the cabinet. It is typical for these cabinets to use a system of rotating wheels with vanes to propel the abrasive, from which the term "wheelabrator" has been adapted for general use. These cabinets also have an abrasive recovery and recycling system, and are capable of very high cleaning rates.

Centrifugal blasting is a most efficient and economical way to prepare pipe for coating in a stationary plant. A series of centrifugal blast wheels housed in a blast enclosure ensures the rotating pipe is cleaned as it travels through the blast machine. There may be a series of one or more blast machines in the line to allow for proper blasting and profile.

If a pipe has heavy mill scale, the first blast machine may be filled with a proper-size shot, since shot helps remove mill scale better than grit does. The second blast machine will then have grit, which leaves the proper profile. If only one blast machine is used, a combination of grit and shot are applied. Centrifugal-blasting equipment works best in a stationary plant set-up, which eliminates the need for a compressor that includes air/blast hoses, abrasive pot, and an attendant.

8.2.3.3. Manual-blasting Technique

Manual blasting should systematically cover the entire surface to be cleaned, with an operator moving the nozzle at fairly constant speeds in straight paths with each successive pass overlapping the preceding one and exposing clean metal without any discolored patches. The standard of blast cleaning should be no more, and certainly no less, than is required by the specification. The nozzle should be held close to right angles (90 degrees) to the surface, but at a slight angle so that the abrasive does not bounce back at the operator.

Some surfaces, such as those with heavy mill-scale layers, are best initially blasted at a shallower angle (e.g., 45 degrees). The operator must be aware that this technique, while efficiently removing the existing coating layer, produces a reduced-surface profile due to the impact angle. Final blasting at right angles to the surface is required for achieving the correct surface profile.

8.2.4. Surface Profile

In addition to cleaning the substrate, abrasive blasting alters it from a smooth surface to a uniformly textured surface. This surface is the result of the sharp, abrasive particles striking the steel at high speed and leaving small impact craters or irregularities. This texture is called "surface profile" or "anchor pattern."

A wellwritten coating specification will include a range of surface profile depths, expressed either in mils or micrometers. For example, a specification may call for a surface profile of 1.5 mils to 3.5 mils (37 μm to 87 μm). Surface profile is important, in that it significantly increases the total surface area and roughness (anchor tooth) to which the coating can adhere. Too low a profile may result in premature coating failure due to lack of adhesion, seen as peeling, blistering, or delamination. Too high a profile may create peaks that will be inadequately covered. This effect is most likely caused when primers are applied but left exposed (without topcoats) for some period of time. This is typically not a problem with most pipeline coating, but it can be an issue for valves and other components that may have multiple layers. Good practice suggests applying at least two coats of a coating system over the blast-cleaned surface to ensure adequate coverage of the surface profile. In general, the greater the surface profile, the better the coating adhesion.

Surface-profile depth can be evaluated by several methods (i.e., comparator and coupons, replica tape, and dial-gauge depth micrometer or profilometer).

8.2.4.1. Surface-profile Coupons

Surface-profile coupons are available in 0.5 mil (12 μm) increments from 1/2 mil to 3 mils (12 μm to 75 μm). The coupons allow for the determination of surface profile through comparison (ASTM D 44179 1, Method A). Other examples of surface-profile coupons are the ISO 8501-3 Comparators, for G-grit and S-shot. ISO 8503 describes two types of surface-profile coupons, Type G for grit abrasives and Type S for shot abrasives. With the assistance of a 5X- (and not to exceed 7X-) lighted magnifier, the profile-reference comparator is placed on the blasted surface to assess the profiles that are nearest the profile of the blasted surface and determine grade.

Five grades may be recorded:

1. Finer than Fine–Any profile assessed as being lower than the limit for fine
2. Fine–Profile equal to segment 1 and up to, but excluding segment 2
3. Medium–Profiles equal to segment 2 and up to, but excluding segment 3
4. Coarse–Profiles equal to segment 3 and up to, but excluding segment 4
5. Coarser than Coarse–Any profile assessed as being greater than the upper limit for coarse

The comparators are accompanied by a card stating the parameters of ISO 8503 Part 1 and Part 2. Comparator assessments are to be reported by the user as being one of the five grades, not as segment numbers.

8.2.4.2. Surface-profile Comparator

The surface-profile comparator consists of a reference disc and a fivepower illuminated magnifier. The disc has five separate leaves, each of which is assigned a number representative of the leaf's profile depth. The reference disc is compared with the surface through the fivepower magnifier. The leaf that most closely approximates the surface roughness is considered to be that surface's pattern. Reference discs are available for sand, grit/steel, or shot abrasives.

8.2.4.3. Replica Tape

Surface profile may be measured with replica tapes. Two types of tape are commonly used (i.e., coarse, for 0.8 mil to 2.0 mil (20 μm to 50 μm) surface profile, and extra-coarse, for 1.5 mil to 4.5 mil (37 μm to 112 μm) surface profile).

A piece of tape with a small square of compressible foam plastic attached to a non-compressible plastic (Mylar) film is applied to the blast-cleaned surface, dull side down. A hard, rounded object (burnishing tool) such as a swizzle stick, is then used to crush the foam to the blast-cleaned surface, causing the foam to form an exact reverse impression (replica) of the actual surface profile. The tape is removed from the surface and an anvil micrometer measures the thickness of the foam and the plastic. The thickness of the Mylar film (2 mils, 50 μm) is subtracted from the micrometer reading, and the result is the depth of the surface profile.

8.2.4.4. Electronic Profilometers

Profilometers are becoming more popular in some areas and in some industries, but have not been accepted everywhere. These electronic meters do offer a good and fast way of determining the profile. If the surface profile (anchor pattern) is found by measuring to be less than was specified, a deeper profile can be achieved by re-blasting with a more aggressive or bigger abrasive, and possibly at greater air pressure. If the surface profile (anchor pattern) is found by measuring to be greater than specified, remedial work may not be possible.

Some profilometers now include electronics that will scan the surface of the blasted surface and provide a printout of the profile showing the number, height, and depth of peaks and valleys. This is important for understanding the total profile. The tape will indicate the height of the profile, but may only indicate a few peaks and not the total number of peaks. These instruments are the newest development for taking profiles.

8.3. Coating Application

Coating application is an important factor in any coating system's performance. The quality and physical properties of a coating material are determined by the manufacturer, but its potential performance can only be reached if the coating is properly applied. Other factors are also important, such as surface preparation, selection of the correct coating for a specific service environment, and inspection.

To become a successful protective coating, the material must be transferred from its primary container to the surface to be protected and must then form a cohesive film with desired properties. The film must be dense, resistant to the passage of moisture and other potentially damaging or corrosive materials, and must dry or cure to its solid state (some coating types do not require cure). The application process plays a significant role in film formation.

8.3.1. Application Methods

Many methods are used to apply protective coatings for pipeline use, including:

- Brush (field and shop)
- Roller (hand or power) or trowel (field and shop)
- Spray (including conventional air spray, airless spray, or some modification of these)
- Powder–Electrostatic (plant)
- Powder–Hand flocking (field and shop)
- Extrusion–Side or crosshead die (plant)
- Wrap–with and without wrapster (field, shop and plant)

Each coating type has preferred ways of application. Many times these methods can be adapted as per the environment and location of the coating. One or more of the following may influence the choice of the method used:

Size and type of job. Bigger jobs are more likely to use more equipment and more sophisticated equipment. The type of job (defined by the specification) determines the required or most suitable application method. New pipe and components are usually coated in a plant or shop. In-service pipe is coated in the field.

Accessibility of areas to be coated. For practical reasons, some projects will restrict the type of application equipment used. A good example of this is in-service pipe exposed in a ditch. Many times there is limited space around the pipe because of ditch size, other pipes in the ditch, or rocks, water, and other obstructions.

Configuration of areas to be coated. Valves and other components can be complex and may be difficult to coat. Pipes and bends can be coated by many methods according to the environment.

Presence of critical areas or surrounding environment that could be damaged by overspray, blasting, fumes, etc. Increasingly, there is public resistance to debris (such as overspray) drifting off the job site. Full containment of the work area may be possible, but the use of brushes and rollers rather than spray equipment may be a more economical solution to the problem.

Type of coating. The type of coating usually dictates the application method(s). Most liquid coatings are designed for application by spray techniques, although brush or roller application can be used when spray application is not possible or for small areas such as repair areas. Powder coatings can be field-applied with proper heating and flocking equipment. Tapes can be applied by hand or wrapping devices. Shrink sleeves require heating devices and rollers, etc.

Availability of skilled workers. This is a critical step that is many times overlooked to save money. All workers should be trained and tested to prove they have the skills to apply the specified type of coating. More sophisticated coatings and equipment require individual applicators to have significant expertise. In many geographical areas, the skill level is simply not available for specifications or the use of such materials.

Budget constraints. If money is not available to pay for more-expensive coatings or application equipment, choice may be limited to simple materials and simple application techniques.

8.3.2. Brush Application

Brush application of coating is the traditional method, although in pipelines it has been largely superseded by spray application techniques. Brush application is slower than other methods and is generally used in the following applications.

- On smaller jobs such as field joints, where application by roller or spray may not be feasible, and for repair or *touch-up* of damaged areas
- For *cutting in* corners or edges of valves and other irregular surfaces
- To achieve good penetration into crevices or pits, especially when applying a primer before the application of the tape or other material
- In critical areas where spray application, if used, may cause damage because of overspray on surrounding surfaces
- For stripe coating of welds, bolts, nuts, edges, flanges, corners, etc.

However, there is no doubt that the rubbing action of a brush can be an additional aid to good adhesion. For this reason, brushing is often preferred for the application of primers and is also recommended for the general application of underwater compositions. Brush application can be advantageous for the application of coatings on surfaces that cannot be fully and properly cleaned. The superior *wetting* action of a brush achieves better contact between the coating and the surface. Brushing also allows working the coating into weld ripples.

A brush of suitable size for the work-at-hand should always be selected; using a small brush on a large area makes it difficult to apply an even coating and slows down the rate of working, while using a large brush on a narrow area can leave sloppy, inaccurate work.

8.3.3. Roller Application

Roller application is of particular value on pipe. Although not as quick as spraying, it is usually quicker than brush application. Another advantage is that it enables a semi-skilled applicator to obtain a reasonable and consistent standard of finish.

There are no particular difficulties in roller application and the technique is soon acquired. For large jobs it is more convenient to work from a bucket than a tray; a perforated grid is placed inside the bucket, the roller being dipped into the coating and then rolled over the grid to remove surplus material and distribute it evenly. With the tray, a reservoir at one end holds the coating; after being charged with coating, the roller is rolled out on the platform of the tray.

The covering material for the rollers may be short-haired carpet-pile fabric, long-haired lambs' wool, or sponge plastic. Selecting the correct length of roller pile for a particular coating is critical to successful application. If necessary, coating manufacturer advice should be taken for the best type of roller material.

The quality of the coating film produced also depends on the roller nap. This term refers to the make-up of the fabric covering (e.g., length and density), and affects the quantity of coating applied and the texture of the applied film.

Rollers must always be cleaned immediately after use. Problems with the use of rollers generally arise from careless handling techniques and equipment maintenance. Rollers are not effective with forcing coating into pitted areas or displacing residual traces of loose dust and dirt from the surface. There is also a tendency for operators to apply a heavy coat at the beginning of a patch, and thinning out the coat to an inadequate thickness before a roller is recharged with coating.

8.3.4. Coating Application by Spray

Generally, spray application is the best method for rapid application of coatings to large pipes and for the uniform application of most coatings. There are two major types of spray-application equipment. With conventional air spray, the coating is atomized by a stream of compressed air and carried to the surface on an air current. Air and coating enter the gun through separate passages (channels), are mixed, and are driven through the air cap in a controlled spray pattern. In airless spray, the coating is atomized without the use of compressed air and is then carried to the surface by the power of the fluid pressure passing through the spray gun. The coating is pumped, under high pressure, to the airless spray gun, where it is forced through a precisely shaped and sized opening (called the orifice or spray tip) as it is being driven to the surface.

Both conventional air spray and airless spray equipment form the basis for modified equipment for special use situations, including plural-component spray, hot spray, electrostatic spray, centrifugal spray, high-volume low-pressure (HVLP) spray, and air-assisted airless spray.

The advantages of the conventional air spray include the spray pattern being adjusted easily to almost any desired fan width and the use of high-quality finishes, such as those for valves and other components that may be above-ground. The disadvantages include the high loss of coating caused by overspray, billowing, and air turbulence created by the compressed air, and reduction of coatings with solvent for proper atomization, resulting in lowered dry-film thickness (DFT) per application.

For airless spray, the advantages include reduced overspray and bounce-back, resulting in material savings; heavier film builds with most coatings; compressed air is unrequired for the atomized coating; a pressure pot is unrequired; equipment is powered by air, electricity, or hydraulics; a faster production rate; and coating is driven into crevices, cracks, and corners. The disadvantages of this method can include fixed, not variable, fan width of an individual spray tip; little control over the quantity of coating applied except by changing tips; and difficulty with coating small, intricate items.

Conventional spray is widely used for high-quality finishes (e.g., car spraying) and a broad range of coating applications. However, it is relatively slow and provides a low film build. High-build coatings can be sprayed with conventional spray equipment, but the material generally requires thinning to pass through the gun at relatively low pressure. Some users continue to use conventional spray equipment because it is less hazardous than other application tools.

There are safety hazards associated with all spray operations, regardless of what types of spray equipment are used. Specific cautions for each type of spray equipment are covered in the respective discussion.

8.3.4.1. Fire and Explosion Hazards

Danger from toxic or fire hazards should always be in the minds of coating inspectors as well as supervisors and workers. Most workers are usually aware of hazards from mechanical equipment, track hoes, ladders, etc. However, they may not realize the tremendous damage that can result from a small quantity of vaporized volatile solvent (an explosion hazard); in addition, they must be made aware of the health dangers inherent to fume and dust exposure.

Workers have been killed by explosions resulting from applying coating in confined places due to toxic hazards. One accident occurred when workers wore proper masks, but the concentration of toxic vapor in the air space was in the explosive range. An extension light bulb broke and ignited the vapor, killing several men.

Flash point is the lowest temperature at which vapors of a flammable solvent will ignite or explode. A surface may be hot enough to volatilize sufficient solvent for a localized danger. Sprays and mists can be very dangerous; even finely atomized metals or dusts may explode when dispersed in air. Coatings and their solvents are sometimes categorized according to flash-point temperature. Low flash-point solvents are those with a flash point below typical storage temperatures (73°F or 23°C), and are the most hazardous to store and/or use. Adequate ventilation is essential for keeping solvent content in the air below the lower explosive limit (LEL). Ventilation also facilitates curing of the coating.

Static electricity may discharge and ignite solvent vapors. This hazard may be reduced by grounding the spray equipment, ensuring that connections are electrically continuous.

8.3.4.2 Breathing Apparatus

Spray finishing creates a certain amount of overspray, hazardous vapors, and toxic fumes. This is true even under ideal conditions and there is no way to avoid it entirely. Anyone who is near a spray-finishing operation should consider some type of respirator or breathing apparatus.

A respirator is a mask worn over the mouth and nose to prevent the inhalation of overspray fumes and vapor. Respirators are necessary for two reasons. First, some sort of respiratory protection is dictated by regulations, such as those formulated by the Occupational Safety and Health Administration (OSHA) and the National Institute for Occupational Safety and Health (NIOSH). Second, even without regulations, common sense determines that inhaling coating overspray and solvent fumes is unhealthy.

Even though a concentration of flammable gas or vapor may be below the LEL, it may be far above the safe limit for breathing. The maximum allowable concentration (MAC) is the amount that must not be exceeded for workers exposed to the hazard during an eight-hour workday. This concentration pertains to vapors, gases, mists, and solids. MACs are published annually (in the United States). These same MACs are often adopted in other countries.

Four primary types of respirators are available to protect the operator.

Air-supplied hood respirator. Hood respirators are designed to cover the entire head and neck area and supply the wearer with clean, dry air at low pressure through a filtered air supply. These respirators protect wearers from heavy concentrations of vapor, fumes, dust, and dirt that might prove harmful to respiratory organs, eyes, ears, and exposed skin. They are used where other types of respirators are impractical and do not provide sufficient protection. The hood respirator provides the most complete means of protection because it offers eye, ear, and skin protection. The continuous supply of dry fresh air prevents misting or fogging in the hood. Air-supplied respirators are commonly required for coating work in confined spaces (such as tanks) and may be mandatory when certain coatings (e.g., those containing isocyanates) are spray-applied.

Air-supplied mask respirator. The air-supplied mask respirator only covers the nose and mouth or it may be full-face, and operates from an external supply of air. It does not provide the degree of protection against splashes, etc., that can be achieved with a hood respirator. If a full-face respirator is not used, eye protection, such as goggles, must also be worn.

Organic vapor cartridge respirator. The organic vapor respirator covers the nose and mouth and is equipped with a replacement cartridge designed to remove organic vapors through chemical absorption. The correct cartridges must be used. Some of these respirators are also designed to remove solid particles from the air before it passes through the chemical cartridge. It is usually used in a finishing operation with standard materials and is not recommended for use in commercial-coating operations. To be effective, there must be a complete seal between the mask and the face. Separate safety goggles or other eye protection must be worn when required. Records of respirator use should be maintained, since the cartridges have a limited life and must be replaced.

Dust Respirator. Dust respirators are sometimes used by sprayers or helpers. These respirators are not effective and are probably illegal. Equipped only with a cartridge (filter fabric) that removes solid particles from the air, these respirators are used during preliminary surface preparation operations like sanding, grinding, or buffing, and are not designed to remove vapors. Separate safety goggles or other eye protection must be worn when required.

8.3.4.3. Personal Protective Equipment

Safety recommendations for proper personal protective equipment (PPE) can be found on the coating manufacturer's material safety data sheets (MSDS). d specified clothing, such as gloves, masks, and long-sleeve shirts, should always be worn.

8.3.4.4. Conventional Spray Equipment

Air-control equipment is any piece of equipment installed between an air compressor and the point of use that modifies the nature of the air stream. Air-control equipment can modify, or control, the volume of air, air pressure, cleanliness of air to the spray gun, and distribution of air to multiple pieces of equipment.

Air-control equipment is often collectively known as the *air transformer* (also called a *filter* or *regulator*). This multipurpose device removes oil, dirt, and moisture from compressed air, regulates, and indicates by gauge, the regulated air pressure, and provides multiple air outlets for spray guns and other air-operated tools.

The principal parts of an air transformer are an air condenser; a filter installed in the air line between the compressor and the point of use to separate oil, water and dirt, and cool the compressed air; and an air regulator, a device to reduce the main line air pressure as it comes from the compressor. Air regulators are available in a wide range of air volumes and pressures, with or without pressure gauges, and in different degrees of sensitivity and accuracy. They have main air inlets and regulated air outlets.

8.3.5. Coating Application by Airless Spray

Airless spray differs from conventional air spray because it does not use compressed air to atomize the coating. Instead, coating is pumped from a container, usually the manufacturer's original cans but sometimes bulk-supply drums (200 L), through a supply line to the airless spray gun. Airless spray operates by forcing coating, at high pressures, through an accurately designed small hole or orifice. As the coating leaves the gun and meets the atmosphere, it expands rapidly. These two factors cause the coating to break up into an extremely fine, very even spray pattern. Air is not used to atomize the coating, hence the airless label.

In airless spray equipment, the material is under high pressure between the pump and the gun, but unlike pressure-feed air spray, the material is not under pressure in the container. Thus, material may be drawn directly from the original container by suction from the pump. Advantages of airless spray include:

- Production rates are increased (faster application). Airless spray applies most types of coatings faster than any other manually operated method of application.
- Because the coating container is not under pressure, the pump can operate from the manufacturer's container.
- Because air is not used for atomization, overspray is much reduced. A degree of coating *drop-out* may occur. This can be reduced by pressure control.
- Blowback is minimized.
- A uniform, thick coating is produced, reducing the number of coats required.
- A very *wet* coating is applied, ensuring good adhesion and flow out.
- Most coatings can be sprayed with very little added thinner. With less solvent, the material dries faster and is less harmful to the environment.

- Coating penetrates better into pits, crevices, recessed areas, and hard-to-reach areas (such as corners).
- The single-hose connection to the gun makes it easier to handle.

8.3.5.1. Airless Spray Safety

Safety precautions for using airless spray are essentially the same as those for conventional air-spray equipment with one very important addition. Airless spray operates by forcing materials at very high pressure through a very small opening. The atomization achieved is so effective that liquids may be passed through a membrane (e.g., human skin) without breaking it. This is the same principle used with the high-pressure devices the military uses instead of hypodermic needles to give military personnel their medical shots.

The accidental injection of coating materials is a very real and present danger. Accidental injection–if untreated–may result in the loss of a limb or may even be fatal. It is advisable when working with or near airless-spray equipment to treat an airless spray gun as though it were a loaded weapon.

Safety authorities (e.g., OSHA in the United States) recognize the danger and require that airless-spray guns carry safety warnings and that they be fitted with a safety spacer at the tip (i.e., the point where coatings leave the gun). The intention of the spacer is to reduce the possibility of injecting coatings or solvents into skin without breaking the surface.

Injection of solvents or other fluids into the skin damages local tissue and may introduce the fluid into the bloodstream. Localized swelling occurs and continues to occur until the pressure is relieved. Proper treatment involves release of pressure and toxic chemicals, generally by cutting open the skin areas affected. The resulting wound may be significant.

An accidental injection is highly unlikely if all safety precautions are observed. However, should a person be accidentally injected, he or she should be taken to a doctor immediately even when an injury seems minor. Delay may cause loss of a finger, an arm or a leg, or even death.

Some additional rules for airless spray safety are:

- A pressurized unit should never be left unattended. Operators should shut the unit off, relieve the pressure, safety-engage the spray gun's trigger, and shut off the power before leaving the area.
- All fluid connections should be high-pressure rated airless-spray fittings, tightened securely, and checked before each use.
- The fluid hose should be electrically grounded to reduce the hazard of static electricity sparks.
- The coating and solvent manufacturers' safety precautions and warnings should be followed.
- Any unsafe condition or practice should be reported to the safety supervisor immediately.

8.3.5.2. Airless Spray Equipment

The most commonly used airless spray system is the direct-supply type. In this system, the pump operates only during spraying. It starts when the spray gun is triggered, stalls out against pressure, and stops when the trigger is released. A typical direct-supply system contains coating supply, pump, filter, hose, and spray gun.

Airless spray pumps. An airless spray pump draws the coating from the container and supplies it under pressure to the rest of the airless-spray system. Most pumps are reciprocating, positive-displacement types that deliver coating under pressure on both the up stroke and down stroke.

Pump volume is rated in gallons per minute (gpm) or liters per minute (L/min) and depends on the coating pump displacement and the number of cycles per minute at which it operates. Pumps used in coating applications deliver from 2.5 to 15 gpm (10 to 60 L/min). Fluid pressure can vary, depending on pump design, from about 800 to 6,500 psi (6 to 45 MPa). The most commonly used airless spray pumps deliver coatings at pressures in the range of 1,500 to 3,500 psi (10 to 24 MPa) and are powered by compressed air.

Pumps can also be powered by electricity or hydraulics.

Pump output pressure is rated in pounds per square inch or Pascals (psi [MPa]) and, in the case of air-powered pumps, depends on the ratio of the air:motor piston area to the paint:pump piston area and the incoming air pressure. In other words, these pumps work on a fixed-ratio multiplication principle providing a fluid coating pressure that is a multiple of the incoming air pressure.

For example, in a 30:1 ratio pump, 80 psi or 5.5 bar (550 kPa) incoming air pressure results in an approximately 2,400 psi (17 MPa) outgoing fluid pressure. Typical ratios are 25:1, 30:1, 45:1, and 65:1. The higher ratio pumps are preferred for spraying high-solids materials and are essential if more than one spray gun is operated from the one pump.

Cold-weather operation and the use of longer hoses also require additional pressure for successful atomization.

Airless spray material containers. In an airless-spray system, the material container is not pressurized. The coating is drawn from the container by the pump, which then pressurizes the coating. Since the material is drawn by suction through the inlet hose, the hose should be armored to prevent its collapse and the subsequent restriction of coating flow.

Airless spray hose and fittings. The fluid hose, depending on its size and the fittings used with airless equipment, must be designed to safely withstand high pressures (up to 7,500 psi or 52 MPa) produced by these systems. They must also be resistant to the materials and solvents that will pass through them. The most common materials used in airless fluid hoses are nylon, Teflon, and polyurethane. All airless hoses should be electrically grounded to prevent static electricity buildup. The airless unit itself should be electrically grounded in hazardous environments, such as in a live gas plant or an enclosed space. Only fittings, swivels, connections, and other parts designed for high-pressure airless applications should be used when working with airless spray equipment.

Airless spray guns. The airless spray gun forces a coating at high pressure through a small orifice at its tip, thus atomizing the coating and shaping the coating into an oval pattern for application to the work piece.

- The two basic types of airless guns are:
- Internally ported: The pressurized coating passes through the gun body before being forced through the orifice.
- Externally ported: The coating is carried to the orifice through a tube on the outside of the gun.

The major components of an airless spray gun include:

- Inlet: Usually a 0.25-in. (6-mm) threaded nipple to which the grounded fluid hose is attached
- Material port: Carries pressurized coating from the inlet to the diffuser
- Safety tip guard: Required by safety regulations, the safety tip is generally colored bright safety orange or red. The safety tip prevents anyone from getting part of his or her body close enough to the orifice to receive an injection of coating.
- Orifice (or spray tip)
- Gasket: Ensures a tight seal between the fluid tip and the diffuser, thus preventing high-pressure leaks
- Diffuser: Helps the efficiency of atomization. The diffuser has a 0.090-in. (2.3-mm) orifice with a bar inside that splits the high-pressure stream of material. The diffuser is also a built-in safety device, designed to break up the high-pressure coating stream should the gun be triggered without a spray tip in position to atomize it.
- Trigger safety: When in the On position, the trigger safety prevents the gun from being discharged, just as a safety on a firearm can prevent accidental discharge. When in the Off position, as shown below, the gun may be triggered and used.

8.3.5.3. Airless Spray Application Technique

Good airless spray technique is much like that for air spray, except:

- The gun is held farther from the work surface.
- Coating thickness is greater, which requires less pattern overlapping.
- There is a more positive action when triggering the gun.

The spray gun should be held 14 to 18 in. (350 to 450 mm) from the work. The distance varies with the covering ability of the coating, type of surface to be sprayed, and desired spray pattern. Also, the gun should be nearly perpendicular to the surface.

The operator should try to maintain an 8 to 12 in. (200 to 300 mm) fan width (width of spray pattern). This makes a good wet spray pattern. If the gun is not perpendicular to the surface and within 14 in. (350 mm), a spray tip with a narrower fan angle should be used to keep the proper fan width.

Runs, sags, or thin coating can result from poor operator technique but these are more often caused by improper tip size. Remember, too, that as the tip wears, the fan angle narrows, giving a wetter coat. Generally, a tip should be replaced after spraying a maximum of 100 to 150 gal (380 to 570 L) of coating.

Move the gun at a constant speed through the entire stroke.

For a wider spray pattern, hold the gun farther from the work or use a tip with a wider fan angle. This applies a thinner coating. To maintain the same coating thickness, also use a larger spray tip and/or move the gun more slowly.

For a heavier coating, make slower strokes with the gun or hold the gun closer to the work. However, this creates a narrower spray pattern; to maintain the same pattern width, change the tip to one with a greater fan angle. Otherwise, use a larger tip for a heavier coat.

When spraying complex shapes or objects with critical areas, work out the best combination of spray-gun movements to get good coverage without excess coating buildup or sags.

Table 8-2 describes some troubleshooting issues that may be used to remedy application defects.

Table 8-2. Troubleshooting Issues in Airless Spray Application.

Spray Pattern Problem	Cause	Correction
Tails, fingering	Inadequate fluid delivery	Increase fluid pressure Change to larger tip orifice
	Fluid not atomizing	Reduce fluid viscosity Clean gun and filter(s) Reduce number of guns using pump Install sapphire insert
	Pulsating fluid delivery	Change to a smaller tip orifice Install pulsation chamber in system or drain existing one Reduce number of guns using pump Increase air supply (volume) to air motor Remove restrictions in system; clean tip screen or filter, if used
	Suction leak	Inspect for siphon hose leak
Hour glass	Inadequate fluid delivery	Same as a through e, above
Uneven spray pattern	Worn spray tip	Replace tip

8.3.5.4. Operation

The airless unit is very simple to operate. Only one adjustment is necessary to control the flow of coating to the gun, and no critical balancing of air and fluid pressure is required to obtain perfect atomizing action. The size of the spray pattern cannot be altered by fluid-pressure adjustment, but is governed by the type of tip fitted and by the viscosity of the material used.

The airless spraying technique differs slightly from conventional spraying. The spray gun must be held at right angles to the surface throughout the stroke, but the distance between the gun and surface should be about 14 to 18 in. (350 to 450 mm). Strokes may be overlapped 50% to obtain coverage, but the speed of stroke is faster than with normal spraying.

Feathering is not possible with the airless method. The trigger must be compressed firmly and completely at the beginning of each stroke and released abruptly and fully at the end of it. The movement of a stroke should commence before the trigger is pulled and carried on after it has been released to ensure that the operative part of the stroke is at a constant speed.

8.3.6. Powder-coating Application

FBE coatings are very popular pipeline coatings with a very successful history of use. The plant-applied FBE coating with the electrostatic process is discussed in Chapter 3. Figure 8-1 shows the electrostatic spray process in an FBE-coating plant.

Figure 8-1. Electrostatic spray process for FBE on heated pipe.

8.3.6.1. Coating by Extrusion

There are several coating systems that use the extrusion method to apply polyolefin coatings to pipe. These coating application methods are discussed in Chapter 3.

The side extrusion method is used for applying polyolefin coatings over a mastic layer or as the outer layers of the three-layer systems. These are applied as the pipe rotates down the coating conveyor.

This process is similar to that of the plant-applied tape coatings. As the pipe rotates down the conveyor, the heated polyolefin sheet is started on the hot primed pipe (usually with FBE for three-layer systems). This causes the sheet to wrap around the pipe in a spiral. Figure 8-2 shows the side extrusion method used on a three-layer coating application.

Figure 8-2. Side extrusion for a three-layer coating system.

Cross-head die extrusion is used to coat over a mastic-type base layer. In this case, the pipe does not rotate, but is pushed through the circular die that extrudes the melted polyolefin onto the pipe. Chapter 3 discusses the process more completely.

8.3.6.2. Wrapping

Tape coatings are applied by hand or by some type of wrapping machine. Hand application is limited, since it does not provide the same tension control and overlap control as a wrapster machine.

Hand wrapping small-diameter pipes (≤ 4 in. diameter) is much easier than hand wrapping larger diameter pipes. Solid-film tape wraps are easier to hand-apply than the mesh-backed tapes, the solid film-backed tape has more issues with soil stress, etc. The backing on some solid-film tape coatings is not strong enough for use of a tape wrapster. The larger the pipe, the more likely there will be soil stress and damage to the tape coating.

The mesh-backed tapes are more suitable for use with a wrapster because of the strength of the backing and its resistance to stretch. This allows for the mesh-backed tapes to be applied with considerable tension, which helps to seal the helix areas at the overlaps and provides additional protection against soil stress. Additional "slip plane" outer wraps can be applied over the mesh-backed tape to help prevent soil stress. Figure 8-3 shows application of a mesh-backed tape with a hand wrapster.

Figure 8-3. Mesh- backed tape being applied with a hand wrapster.

Plant-applied tapes are applied by a similar method as that of the side extrusion of polyolefin. The prepared and primed pipe rotates down a conveyor as the tensioned tape is wrapped around the pipe. This is a multilayer process that allows for setting the proper tension and overlap. Figure 8-4 shows plant application of a three-layer solid film-backed tape.

Figure 8-4. Plant application of three layers of solid film-backed tape on water pipe.

More information about the differences between solid film-backed tapes and mesh-backed tapes can be found in Chapter 3.

8.4. Test Instruments

Nondestructive test instruments, when used properly, do not destroy the coating on which they are used. Most types of dry-film thickness (DFT) gauges are nondestructive. Essential quality control instruments include:

- Wet-film thickness gauges
- Dry-film thickness gauges, including pull-off magnetic gauges and fixed-probe magnetic gauges
- Holiday detectors, including low-voltage DC (wet sponge) detector, high-voltage DC detector, and high-voltage AC detector.

8.4.1. Wet-film Thickness Checks

Measurement of wet coating film thickness provides a useful guide to ensure that a correct and even film thickness is being applied to the article being coated. Use of a wet-film comb at this stage of a coating application operation helps prevent rejected work which, in itself, is time-consuming and costly.

The most common instrument for measuring wet-film thickness (WFT) is the comb gauge. WFT measurements should be made as soon as possible after coating application. The test gauge may leave marks in the coated surface that may adversely affect coating integrity. These marks should be immediately over-coated with fresh coating to avoid pinholes. Guidance for the use of the WFT gauge may be provided by the specification.

8.4.2. Wet-film Thickness Gauge

An essential companion to any instrument used to measure DFT is the WFT gauge. Using knowledge of the volume solids content of the coating, the applicator can calculate the WFT required to produce the desired DFT. Typical WFT gauges are shown in Figure 8-5.

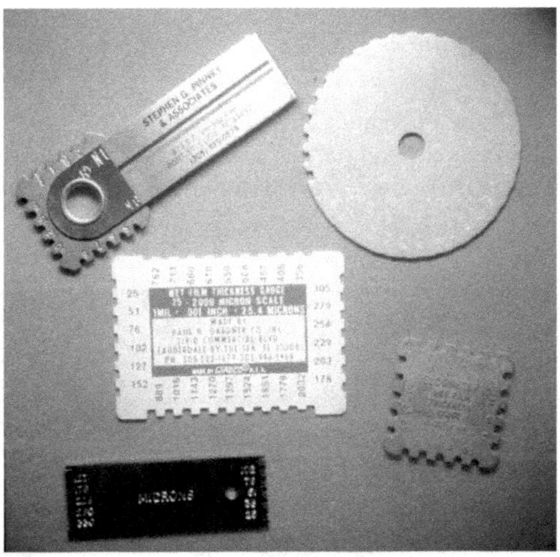

Figure 8-5. Wet-film thickness gauges.

WFT gauges consist of a series of graduated teeth lying between two outer teeth. WFT gauges with different scales (e.g., mils, μm) are available. The gauge is pushed firmly into the wet coating film so that the outermost teeth make contact with the substrate or previously coated surface. The gauge must be at right angles to the surface.

The gauge is removed and the teeth examined. Some of the heads of the teeth are coated while others are clean. The true WFT lies between the last tooth with a coating and the next (higher) tooth that is uncoated. The reported WFT is that of the last wet or coated tooth on the gauge.

8.4.3. Dry-film Thickness Checks

DFT is normally measured with an electronic or magnetic gauge. The instrument must be verified if accurate measurements are to be made. To obtain best results, it is required to:

- Verify the instrument on steel with a surface profile matching the profile of the coated surface being measured.
- Verify the instrument in the expected range of DFT to be measured. (Type 1 uses verification blocks and Type 2 uses shims for verification).
- Calibrate the test equipment and provide the calibration certificate with the new instrument. The instrument should be returned to the manufacturer for re-calibration.
- When a particular standard or method is required by the coating specification, use that method.

8.4.4. Magnetic DFT Gauges

Common types of simple DFT measurement instruments include pull-off magnetic DFT gauges, which use a calibrated spring to pull a small permanent magnet from the coated surface (Figure 8-6), and magnetic constant-pressure DFT gauges, which depend on changes in magnetic flux within the probe (Figure 8-7).

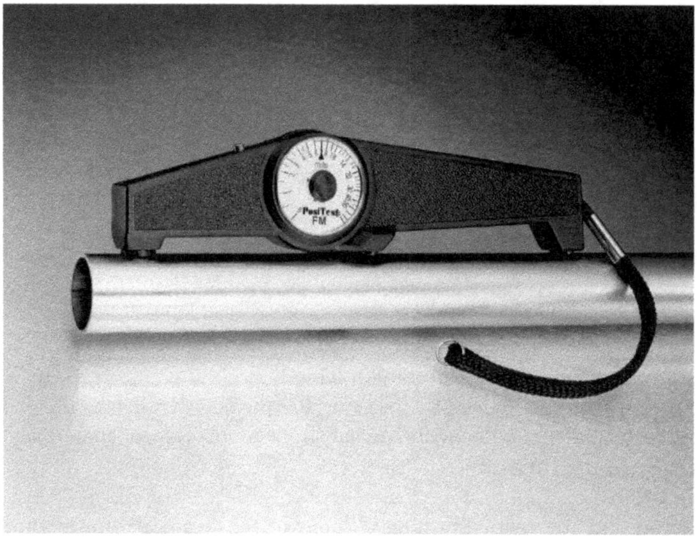

Figure 8-6. Dial-type magnetic pull-off DFT gauge.

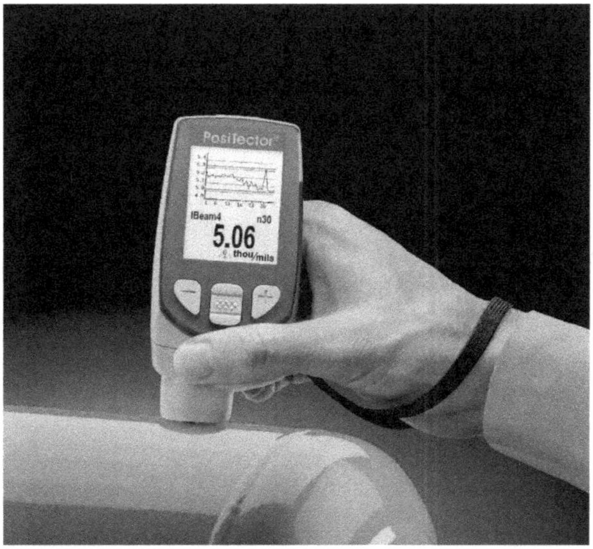

Figure 8-7. Magnetic constant-pressure DFT gauge.

Some care must be taken with all instruments using magnetic probes. An exposed magnet may attract any nearby loose iron and steel shot or grit particles. The magnet should be cleaned of any contaminants that might alter its reading. It is important that the spot on the surface being measured is clean. Otherwise, the reading may not indicate the gap between the surface and magnet as it is designed to do. The gap may, in fact, be made up of rust, residues of abrasive blasting, or other impurities, which could adversely affect the reading. Careful inspection of the surface for cleanliness, both before coating application and between each coat, is important.

If the instrument is used on tacky films, the reading may indicate a thickness less than that of the coating. That is because the film itself will hold the magnet to the surface beyond the point when it would otherwise have pulled away. If used on a soft coating, the tip may depress the coating, causing a thin reading. Vibration of the substrate might cause the magnet to pop off the surface before it otherwise normally would, giving a high thickness reading. Magnetic instruments are also likely to be affected by magnetic fields close to edges. If possible, the instrument should not be used closer than 1 in. (25 mm) from the edge of the surface.

When using dial-type pull-off gauges it is easy to continue to turn the dial after the magnet has lifted from the surface, giving an incorrect reading. New versions of some instruments solve this problem with an automated mechanism that turns the dial at a fixed rate and stops when the magnet lifts from the coated surface.

Thickness measurements should be made after the application of each coat in a multicoated system. However, with nondestructive gauges, it is difficult to determine the exact thickness of each individual successive coat after the first coat. Total DFT measurements for the coating film must be made.

The DFT of individual coats can be estimated by calculation. The second coat, for example, can be determined by subtracting the average thickness of the first coat from the total measured thickness. The DFT value will be an estimate only, because it is unlikely that the second set of measurements will be taken in the same position as those of the first coat and the first-coat DFT may not be truly represented.

Thickness measurements are made to ensure the specification is being met. Obviously, an inspector cannot measure the DFT for every square centimeter of coated surface. Inspectors, therefore, must be able to use some standard or agreed-upon method for measurements. Resulting values should be taken to represent the DFT of the coating film.

8.4.4.1. DFT Measurements with Magnetic Gauges

Various standards define methods of measuring DFT. ASTM D1186 and SSPC-PA 2 define similar methods for verifying magnetic-type DFT gauges and for making DFT measurements of nonmagnetic coatings over a ferrous magnetic metal surface. Of the two specifications, SSPC-PA 2 will be explored here.

Other standards may be required by a particular specification, and the inspector should be careful to use the defined method. If no standard is required, then it would be good practice to choose a standard method, such as SSPC-PA 2, and work within guidelines defined by consensus within the coatings industry. Alternatively, the inspector can propose a method for verification and measurement frequency. The various parties concerned should agree on a method at the pre-job conference.

The requirements of SSPC-PA are as follows: Five spot measurements (average of three readings) for each 100 ft^2 (10 m^2) measured. Individual readings are not subjected to these rules but are included in the average for a spot measurement. The average of five spot measurements (i.e., 15 individual measurements) should be no more and no less than the specified coating thickness range. No single spot measurement can be less than 80% of the specified thickness and no more than 120%.

Minimum thickness. The average of the spot measurements for each 100 ft^2 (10 m^2) area shall not be less than the specified minimum thickness. No single spot measurement in any 100 ft^2 (10 m^2) area shall be less than 80% of the specified minimum thickness. Any gauge reading may be under run by a greater amount. If the average of the spot measurements for a given 100 ft^2 (10 m^2) area meets or exceeds the specified minimum thickness, but one or more spot measurements are less than 80% of the specified minimum thickness, additional measurements may be made to define the nonconforming area.

Maximum thickness. The average of the spot measurements for each 100 ft^2 (10 m^2) area shall not be more than the specified maximum thickness. No single spot measurement in any 100 ft^2 (10 m^2) area shall be more than 120% of the specified maximum thickness. Any gauge reading may overrun by a greater amount. If the average of the spot measurements for a given 100 ft^2 (10 m^2) area meets or falls below the specified maximum thickness, but one or more spot measurements is more than 120% of the specified maximum thickness, additional measurements may be made to define the nonconforming area. Manufacturers' literature may be consulted to determine if higher maximum thickness readings are allowable under specific circumstances.

Area measured. For areas under 1,000 ft^2 (100 m^2), randomly select and measure three 100 ft^2 (10 m^2) areas. If the DFT in those areas complies with the specified limits, proceed. If not, make more measurements to define the nonconforming area, then begin again. For areas over 1,000 ft^2 (100 m^2), measure first 1,000 ft^2 (100 m^2) as above and, provided the DFT is OK, for each additional 1,000 ft^2 (100 m^2) randomly select and measure one 1,000 ft^2 (100 m^2) area.

All of these definitions and procedures may, according to the standard, be varied by agreement.

As with all specified standards, inspectors should take the time to become thoroughly familiar with this specification. SSPC describes verification techniques for using two methods and defines DFT gauges in two categories to correspond to the verification techniques.

Another standard, often referred to in specifications, is ASTM D 1186. This standard also has two verification methods, Method A and Method B, which are related to the type of instrument used. Method A uses coated metal shims (such as NIST) to verify pull-off magnetic gauges. Method B uses nonmagnetic shims, placed on the surface to be coated, to verify magnetic flux (i.e., constant-pressure probe) gauges. Like SSPC-PA 2, this standard measures the thickness of nonmagnetic coatings applied to ferrous metal substrates.

Whichever standard method is used, recording the correct number of measurements is important. The inspector may use a table similar to that shown in Figure 8-8 to ensure that all relevant measurements and calculations have been made.

First Coat

Specified DFT Min:__________ Max.__________

Location:______________________________

Spots->	1	2	3	4	5
1					
2					
3					
Avg.					

Overall Average DFT:__________

Minimum DFT:_______ Maximum DFT:_______

Figure 8-8. Typical form for documenting DFT-thickness measurements.

It is helpful, when nonconforming areas are found, to mark those areas where the coating is found to be either too thick or too thin. This can be done by applying a contrasting color coat of the same coating, but in no case should any marks be made that could damage the coating. The method of marking coating deficiencies should be agreed during the pre-job conference.

8.4.4.2 Magnetic Pull-off DFT Gauge

The magnetic pull-off DFT gauge is a simple mechanical tool that works by magnetic attraction to a ferrous surface. The force of attraction is reduced by the presence of a nonmagnetic barrier between the permanent magnet and the surface. Measurement of the force required to pull the magnet away from the surface is equated with the thickness of the film.

The gauge is used for nondestructive measurement of the DFT of nonmagnetic coatings on a ferrous metal substrate. Magnetic pull-off gauges do not depend on batteries or any other source of power and are therefore considered by many users to be more reliable in the hands of inexperienced users. They are also used in hazardous environments when non-spark instruments are required. They are widely used despite the frequent appearance of more sophisticated and more accurate electronic gauges. They are portable, simple to operate, and inexpensive.

Measurements are made in accordance with the specification or referenced standard. If no specific method is mentioned in the contract documents, this issue should be discussed at the pre-job meeting and a suitable method agreed upon. It is always better to follow an industry standard whenever possible.

Place the gauge on a clean, dry, and cured coated surface. Do not use on soft or tacky films. Rotate the dial all the way to the highest reading on the gauge and then lift the counterbalance so that the magnetic tip contacts the coated surface. Slowly rotate the dial (increasing spring tension) at a constant speed until the magnet pops up from the coated ferrous surface. Where possible, the gauge should be mechanically stabilized by pressure from the operator's hand. Keep the magnetic tip free of magnetic particles and other residues. Do not use the gauge within 1 in. (2.5 cm) of an edge, on or near vibrating equipment, or on metal being welded (the unit may be demagnetized). On circular pieces, locate the gauge along the longitudinal axis.

The instrument must be verified if reliable measurements are to be made. There are different ways to verify magnetic-DFT gauge. If a standard (such as SSPC-PA 2) is specified, the verification method should be that defined by the standard.

An alternative system sometimes used is to obtain a small sample of steel, approximately 6 x 4 in. (15 x 10 cm), and have this blast-cleaned at the start of a project. This panel can then act both as a reference panel for the surface profile agreed on, and a verification panel for verifying DFT measurements.

The accuracy of mechanical gauges such as these is no better than plus or minus 10% in use. With extreme care (e.g., in laboratory conditions), the accuracy may be improved, but operation of the gauge depends on the repetition of the inspector's method of use. Measurements are likely to be affected by the speed of movement of the dial, the proximity of edges and curvature on the structure, and the orientation of the gauge.

The condition of the magnetic probe should be visually checked at frequent intervals, as the gauge tends to attract metal particles. Once the gauge is attached to the probe, these particles change the measurements significantly. Verification checks should also be made at intervals throughout the working day. As with many other gauges, any rogue measurements should be checked. It is not unusual to find an occasional measurement that is quite different from those around the same location. If a measurement cannot be repeated, it should be discarded as invalid and an alternative measurement taken.

8.4.5. Constant-pressure Probe DFT Gauge

Constant-pressure probe DFT gauges are nondestructive instruments that measure DFT of nonmagnetic coatings over ferrous-metal substrates. They determine coating thickness by measuring changes in the magnetic flux within the instrument probe or in the instrument circuitry. The instrument probe must remain in contact with the coating at all times during measurement.

Constant-pressure probe instruments may have fixed integrated probes or separate probes. In each case, the probes are placed against the coated surface and held against the surface while a measurement is made. Probes may be magnetic or electromagnetic. Coating thickness is displayed on the meter or instrument scale. Manufacturers of fixed-probe constant-pressure gauges recommend different methods of calibration or adjustment. Some provide built-in self-calibration routines or can revert to a factory-standard calibration. Some modern gauges have many alternative methods for calibration, each of which is likely to lead to variation in results when measurements are made. Any gauge should be calibrated according to the manufacturer's instructions and/or in accordance with an agreed-upon procedure. Figure 8-9 shows a constant-pressure probe being calibrated with plastic shims.

Figure 8-9. Constant-pressure probe gauge calibrated with plastic shims.

Though some call this step "calibration," it is really "verification" with the instrument reading within the limits of the instrument. Calibration is typically on done at the manufacturer or at a testing lab that has the essential equipment. Verification is performed in the field or plant to determine the settings of the instrumentation and whether adjustments should be made (if possible) to the instrument.

Verification using NIST standards. This procedure follows SSPC-PA 2 for Type-I gauges. First, standardize (check verification of) the gauge by measuring the NIST-test standards within the coating DFT ranges to be measured in the field. If any deviation (+ or –) occurs, the gauge can be physically adjusted until it is accurate, or a verification factor can be noted. This factor is then added to, or subtracted from, any DFT measurements made, as appropriate. Second, measure the blast profile of the steel to be coated and record this data. This measurement establishes an imaginary magnetic baseline in the blast profile. This imaginary line is called the base metal reading (BMR) and is to be deducted from any DFT reading taken on this surface later. The BMR should be a small factor, usually 8 to 20 μm (0.3 to 0.8 mils), but it could be outside this range. When DFT measurements are made, there are two potential corrections. The first is the verification factor, which may be added to or subtracted from the measurement. The second is the BMR, which is subtracted from the measurement.

Verification using nonmagnetic shims. Constant-pressure probe (SSPC-PA 2 Type-II) gauges are generally verified using plastic shims with thicknesses verified with a micrometer. Verification should be made in an area free of magnetic fields (i.e., away from welding equipment, generators, or power lines).

Select shims in the range of expected coating thickness. For example, if the coating DFT is expected to be about 200 μm (8.0 mils), calibrate the unit using a shim as close to 200 μm (8.0 mils) as possible. Some electronic gauges require verification over a range, using either bare steel or a very thin shim at one end of the range and a shim of greater thickness than the coating to be measured at the other end of the range. The accuracy of the gauge after verification should always be verified close to the thickness to be measured.

Place shims on a bare section of the structure to be coated after the specified surface preparation has been completed. Alternately, place the shims on a bare steel plate at least 3 x 3 x 0.125 in. (7.6 x 7.6 x 0.32 cm), free of mill scale and rust. Note that if the calibration surface does not have a profile and the coated surface does, an adjustment should be made to allow for the profile. This typically takes the form of 12 to 20 μm (0.5 to 0.8 mils) that must be subtracted from the measurements made.

One system sometimes used is to obtain a small sample of steel approximately 6 x 4 in. (15 x 10 cm) and have this blast-cleaned at the start of a project. This panel can then act as a *reference* panel for the surface profile agreed on and as a calibration panel to check DFT-gauge measurements. The panel should be of similar material (i.e., steel alloy) and similar thickness to that of the structure being coated.

Avoid excessive pressure that could bend a shim and indent it or impress the peaks of the blasted surface into the contact surface of the shim. Plastic shims used for verification are not made from precision material. Their thickness should be verified with a micrometer.

As with any other DFT gauge, re-verification may be required whenever results appear to be inconsistent or erratic. Battery-powered units may give erratic results as the battery weakens with use.

Fixed-probe gauges are generally more accurate than mechanical gauges, with accuracy around 3% or better. They need a source of electrical power (battery) and are not, therefore, intrinsically safe.

In the United States, smooth verification shims produced by NIST are commonly used. If gauges are verified on a smooth surface, and then used to measure coating thickness over a rough, blast cleaned surface, an adjustment must be made to ensure accuracy.

Experiments have shown that gauges verified on a smooth surface, then used on a grit blasted surface profile of 50 μm (2 mils), measure more than the true DFT by about 25 μm (1 mil) on a coating thickness of 250 μm (10 mils). The allowance in this case would be to subtract 25 μm (1 mil) from every measurement made to obtain a more accurate measurement of DFT.

8.5. Holiday Detection

Holiday detectors are used to detect holidays or pinholes in the coating. General types of holiday detectors include low-voltage direct current, high-voltage direct current (HVDC), and high-voltage alternating current (HVAC) (not normally used on pipelines and will not be discussed).

Holiday testing is performed to find nicks, pinholes, and other defects or discontinuities in the film. Correction of coating defects is especially important for pipelines intended for immersion or burial. The specifications should indicate at what point in the job holiday testing is done. For plant-applied coatings (FBE, two- and three-layer, coal tar, tapes, etc.) the holiday detection is typically performed on the exit rack where the coating is inspected for thickness and other issues. This holiday detection is performed to determine if surface preparation and coating application is correct. Repairs can be made before the pipe is shipped, so these holiday are separated from shipping damage.

The pipe is detected after shipping, handling and storage with the final time just as the pipe is being placed in the ditch or water. Holidays in coatings should be repaired. The type of repair is critical and should only be performed as per the specification and manufacturer's recommendation. The coating should then be tested again, after the repair, to ensure that repairs were successful.

8.5.1. Low-voltage (Wet-sponge) Holiday Detector

This holiday detector is a sensitive, low-voltage (wet-sponge) electronic device powered by a battery with output voltages ranging from 5 to 120 V (DC), depending upon the equipment manufacturer's circuit design. The detector consists of a portable battery-powered electronic instrument, nonconductive handle with clamps (to hold sponge), open-cell sponge (cellulose), and ground wires. The instrument is generally housed in a plastic case with an OFF/ON switch and a socket for headphones. Some low-voltage holiday detectors are fixed at a specific voltage, while others may have a test voltage selected. Some common voltages used are 9, 67.5, 90, and 120 V. Different results are obtained with each voltage, so selecting the proper voltage is important. Ideally, the instrument to be used and its voltage should be specified.

This type of instrument may be used to locate holidays in nonconductive coatings applied to a conductive substrate. According to NACE Standard SP0188, the low-voltage DC detector is generally used on coating films that are less than 500 μm (20 mils) thick. The instrument will still locate defects in coatings thicker than 500 μm (20 mils) and is preferred by some users because it cannot easily damage the coating film tested.

The ground cable is attached directly to the conductive substrate for positive electrical contact. For coated steel, connect directly to the bare metal. This method is not usually not used on coated pipelines, but it may be used in some cases or on components such as valves because of the irregular shape.

The sponge is saturated with a solution consisting of tap water (not distilled water) and a low wetting agent, combined at a ratio of 1 fluid oz wetting agent to 1 gallon U.S. water (7.5 ml wetting agent to 1liter water). This represents a ratio of 1 part wetting agent to 128 parts water. The sponge is wetted sufficiently to barely avoid dripping the solution while the sponge is moved over the coating.

Contact a bare spot on the conductive substrate with the wetted sponge to verify that the instrument is properly grounded. This procedure should be repeated periodically during the test.

With the ground wire attached to the substrate, wipe the coated surface with the wetted sponge at a maximum rate of 1 linear ft (30 cm) per second. Avoid using excess water in the sponge because the rundown may complete the circuit across the coating surface to a flaw located several centimeters away and give false readings. Use a double stroke of the sponge electrode over each area. This ensures better inspection coverage. When a holiday is found, the unit will emit an audible tone.

The detector is factory-calibrated and calibration in the field is not generally necessary. Typical factory calibration is set at 700 μA (10%) of current flow to complete the circuit for the audible signal to indicate a coating holiday on metal substrates.

Wet-sponge holiday detectors are portable and easy to operate. They can be used on coatings up to 500 μm (20 mils) thick with reliability. They are nondestructive and do not harm the coating as the test is made. The test procedure can be slow in operation, taking many hours to fully test coatings in a large vessel. The units are generally not intrinsically safe and cannot therefore be used in a hazardous environment.

8.5.2. High-voltage Pulse-type DC Holiday Detector

High-voltage pulse-type holiday detectors generally have a voltage output range of about 900 to 15,000 V and in some cases may range as high as 40,000 V. They are designed for locating holidays in nonconductive coatings applied over a conductive substrate. Generally, these devices are used on protective coating films ranging in thickness from 300 to 4,000 μm (12 to 160 mils).

The detector consists of a source of electrical energy, such as a battery or high-voltage coil, an exploring electrode, and a ground connection from the detector to the coated substrate. The electrode is passed over the surface. A spark will arc through the air gap or coating to the substrate at any holidays, voids, or discontinuities, simultaneously causing the detector to emit an audible sound.

The ground wire should be connected directly to the metal structure where possible. If direct contact is not possible, the high-voltage holiday detector may be used with a trailing ground wire, provided the structure to be tested is also connected to the ground. This connection may be achieved with direct contact (as when a pipe lays on wet soil) or by fixing a ground wire and spike at some point between the ground and the structure.

Set the voltage as specified or as shown in a referenced standard. If no guidelines are provided, a rule of thumb in industry in the United States is to use a voltage setting of 4 V/µm (100 V/mil) of coating thickness. In Europe, the rule of thumb most often used is slightly different, i.e., 5 V/µm (125 V/mil) of coating thickness.

An alternate method is to make a pinhole (or identify another type of defect, e.g., low DFT) in the coating to the substrate, and set the voltage at the lowest available setting on the unit. Increase the voltage until it is sufficiently high to create a spark at the holiday. Use that setting to inspect the particular coating.

When the voltage is set too high, the coating may be damaged. The same damage might be incurred if the coating is tested before it has released all or most of its solvent content. Once a spark has been generated through the coating to the substrate, a specific holiday exists through the coating, even if it had not been a pinhole or break in the coating before the test was performed.

In using the instrument, move the electrode at a rate of about 1 ft/s (0.3 m/s) in a single pass (according to NACE Standard SP0188). Moving the probe too fast may miss a void; moving the probe too slowly may create damage at thin spots or prove to be more searching than was intended by the specifier.

The accuracy of the instrument can be tested with a dedicated voltmeter connected between the probe and the ground connector. The instrument must be specific to the type of holiday detector, since the pulse characteristics of the signal have to be taken into account. For most users it may be best to send the unit back to the manufacturer for calibration.

Most high-voltage holiday detectors have a wide range of electrodes available for different uses:

- Flat-section rolling springs are used to test pipeline coatings.
- Smooth neoprene flaps (impregnated with conductive carbon) are used for thin-film coatings such as fusion-bonded epoxy.
- Copper-bronze bristle brushes are commonly used on glass-reinforced plastic (GRP) coatings.

High-voltage holiday detectors generate significant electrical energy. While the voltage is not sufficient to kill the operator even at maximum output, it is certainly a shock to the system and may lead to a consequent mishap (such as the operator falling from a scaffold). Operators should wear protective equipment (such as rubber boots) and should not operate the equipment in wet or damp conditions. The unit will give false indications of holidays if used on a wet surface.

The unit is not intrinsically safe and may lead to an explosion if used in an explosive atmosphere. Most holiday detectors provide a constant (low-level) signal, indicating to the operator that the unit is switched on and working. If the unit does not come on, or if the operating signal does not sound, the battery may be dead or weak. The operator should replace or recharge the battery.

If the unit does not spark or emit a sound when the electrode touches the ground, the unit may need repair.

High-voltage holiday detectors are more searching than the low-voltage type. They not only detect any holidays or pinholes that penetrate to the substrate, but can also find defects such as areas of low film thickness or voids hidden within the coating.

This is the reason holiday detectors should not be used on excavated coated pipelines. All coatings absorb some water, so there will be more of a chance to burn through a good coating than there is for finding actual holidays. Some companies or standards may call for using holiday detectors on coated in-service buried or submersed pipe when it is exposed, but this will cause damage to an existing coating in many cases.

References:

Norsworthy, R. "Pipeline Coating Tutorial". Banff Corrosion Workshop, April 2007

Norsworthy, R.; D'Ambrosio, D; Quinn, M.; "Developing Selection Criteria for Field Applied Pipeline Coatings"; CORROSION 2015,

Tator, K.; "Laboratory Testing of Pipeline Exterior Coatings", CORROSION 2006, Paper 06044.

9 Inspection of Buried Pipeline Coatings

9.1. Importance of Coating Inspection

On a well-coated pipeline, missing or damaged coating can lead to an increased CP current demand. If there is no or insufficient CP, corrosion can occur on a steel surface exposed to the environment. Meeting NACE SP0169 CP criteria [NACE, 2013] results in adequate protection of the exposed metal surfaces if those surfaces are not electrically shielded by coatings or other materials. Holidays on pipelines may be detected using various indirect surveys. These surveys evaluate or detect changes in the CP current, voltage potential, or levels of current distribution.

Based on indirect-inspection techniques, if coating holidays are identified, one solution (especially for pipelines that cannot be internally inspected) is to excavate the pipe and visually examine the pipe coating. Anytime a pipe is excavated, the condition of the pipe and the pipe coating should be documented. Regulatory requirements in many countries require this. Best practice is to always document, gather data, and photograph all pipeline exposures.

The ECDA process developed by NACE International [NACE, 2002a] provides the operator with a method for determining where coating defects are located. This method was developed for pipelines that cannot be internally inspected, but it can also be used on other pipelines to provide information about the coating condition. Remember, these methods do not locate external corrosion, but lead the operator to where potential problems may exist.

Several types of internal ILI tools give a picture of the pipe wall and any corrosion or damage that has occurred throughout the service life. Some EMAT tools now also locate missing or disbonded coating areas on a pipeline. These tools allow the operator to be proactive, instead of reactive to potential CP shielding issues and corrosion. [Norsworthy, Jurgk, Heinks, Grillenberger; 2012]

9.2. The ECDA Standard—NACE SP0502

ECDA was developed to help operators determine where potential corrosion may exist on a pipeline where ILI tools cannot be used. The ECDA process uses above-ground techniques to locate areas where coating defects exist. Once a line is exposed, the coating and pipe condition can be determined. The ECDA process may or may not be applicable to some pipelines: This must be studied before the process is implemented. ECDA is only applicable to onshore ferrous pipelines.

The four-step process of ECDA integrates several bits of information that are applied to a method for determining what areas need to be exposed and evaluated. The exposure and evaluation of the area then leads to a direction for future ECDA evaluations. The purpose is for the operator to identify where areas of external corrosion may exist. Each bit of information helps the operator to understand and address areas of concern before there is a failure or before significant corrosion occurs. ILI and pressure tests find corrosion after the fact.

ECDA locates areas where inadequate CP is being provided. The survey techniques basically locate where the current is entering the pipe, and therefore it finds areas where the coating is damaged or missing. For this reason, ECDA is not applicable to poorly coated or bare pipe. There is a special section in the NACE standard that covers these systems.

This process does not measure wall loss or find corrosion that is present. Since these surveys are simply finding where the current is going to the pipe, then the pipe is being protected. The reasons for exposing the pipe are for inspecting the coating defects and deterioration. This evaluation may lead to disbonded CP shielding coating where corrosion is normally an issue. There may also be corrosion that occurred before adequate CP was applied (usually on lines over 40 years old), which needs attention.

9.2.1. ECDA—Step One

Step one of ECDA is to complete a pre-assessment of historical, construction, operations, and maintenance records. The accuracy of these records and information is critical to the success of this process. All departments of the company should be included to ensure that all information and experiences are discovered and properly used.

The data collected from all departments are compiled and organized for proper decisions about what areas are or are not potential risk areas that need to be evaluated. Past ECDA surveys, if available, are excellent tools for making these decisions about risk management and integrity. Data are organized into five categories:

1. pipe-related
2. construction-related
3. soils- and environment-related
4. corrosion protection
5. pipeline operations considerations

Each category is important to the integrity of the pipeline and the amount and accuracy of the data provided in step one. Each tool or survey used will provide only part of the information, but with proper interpretation, use of these categories will lead to a better understanding of the pipeline system.

One must remember there are many things that invalidate information. Some of these are disbonded coating that is CP shielding, roads, low soil moisture, and other shielding structures such as rocks, concrete, and plastic sheets.

A variety of tables and guidelines are available for operators to use to help determine what areas to survey. Only specific sections of the entire pipeline system are surveyed. From these surveys and the interpretation of the direct-assessment data, operators can decide whether to include more of the system.

9.2.2. ECDA—Step Two

Above-ground indirect inspection survey methods locate small data variations over chosen sections. Remember that these surveys do not locate corrosion, but do indicate where current is entering the pipe, which is normally where coating holidays or deterioration has occurred. In current industrial practice, none of the above-ground techniques (including close interval survey (CIS), direct current voltage gradient (DCVG), Pearson, Current Attenuation, etc.) are meant to be used to assess corrosion rates. These techniques cannot measure corrosion potentials under disbonded coatings where CP is shielded. [Song, Sridhar, 2007]

Typically used indirect surveys for ECDA are direct-current voltage gradient (DCVG) surveys, alternating-current voltage gradient (ACVG) surveys, close-interval surveys (CIS), and alternating-current attenuation surveys. Other complementary surveys for the above indirect surveys include pipeline locating, pipe-depth surveys, soil-resistivity measurements, side-drain surveys, and global-positioning system (GPS) surveys.

Several supporting NACE standard practices should be incorporated into the ECDA process, such as:

- NACE SP0169 (latest revision), Control of External Corrosion on Underground or Submerged Metallic Piping Systems
- NACE SP0177 (latest revision), Mitigation of Alternating Current and Lightning Effects on Metallic Structures and Corrosion Control Systems
- NACE Standard TM0497 (latest revision), Measurement Techniques Related to Criteria for Cathodic Protection on Underground or Submerged Metallic Piping Systems

9.2.2.1. Close-interval Potential Survey

CIS provides information about the performance of the CP system. This survey method indicates larger areas of coating defects or deterioration, but does not indicate small holidays and defects. Other advantages of CIS are that it will detect direct-current interference, shorted casings and shorts to other metal structures, geological-current shielding (such as large rocks), and verify whether electrical isolation is working or not.

These surveys are run with two balanced reference cells attached to poles. These are attached to the data recorder with appropriate leads. Then a small gauge wire is attached to the test lead of the pipeline and the technician starts walking over the located and flagged pipeline. The wire attached to the test lead is unrolled as the technician walks the line. Each reference cell is alternately placed over the pipe line at pre-determined spacing (such as one meter), so the potentials are recorded on the data recorder.

This information can then be graphed to provide the operator with information about areas of inadequate CP or large areas of coating defects or deterioration.

9.2.2.2. Direct-current Voltage Gradient

DCVG locates holidays (large and small) where current is being picked up by the exposed pipeline. The experience of the technician is critical to collecting the correct data and location of the holidays. Unlike CIS, DCVG will locate and somewhat "size" the holidays.

An analog voltmeter that is strapped on the technician with the appropriate cables is connected to the meter and electrodes. The electrodes are placed on probes and then filled with water. An interrupter is installed in the rectifier that influences the survey area. The cycle of the interrupter is fast compared to CIS. Typical cycles are 1/3 s ON and 2/3 s off, which allows a quick deflection of the analog meter. These meters have impedance adjustments that allow for the deflection of the needle to be 1 mV or less. As the technician walks the located pipeline, the two probes are placed on the ground. The needle will also deflect in both positive and negative directions, which helps determine the direction of current in the soil and with the proper placement of the probes, helps determine the location and size of the holiday.

9.2.2.3. Alternating-current Voltage Gradient

The ACVG method is similar to the DCVG method, but uses an AC signal to locate the coating holidays. This signal is induced by a low-frequency transmitter that is connected to the pipeline. Holidays are located when there is a change in the signal strength. The shape of the gradient field indicates the type of coating damage. Operator experience is critical, since the correct interpretation of the data is what provides an operator with information about whether to dig or not.

Once again, two probes are used as the technician walks over the located pipeline. Placement of the probes define where a holiday is on the pipeline. The AC-powered signal generator (with specific Hz AC output) is connected to the pipeline and to a ground. The positive and negative cables on the rectifier can be disconnected and used as the connection points of the signal-generating unit since the anodes of the ground bed can be used as the ground for the survey. The negative lead, of course, is attached to the pipe.

The hand-held receiver unit is tuned into the appropriate setting according to the pipeline system being surveyed. Typically, the technician will take the signals at every 3 meters, but will stop and record more data at sights that indicate a coating issue.

9.2.2.4. Evaluation of Indirect Inspections

Once the above-ground surveys and supporting documents have been completed, the information must be evaluated to determine the inspection areas. Priority is given to areas where two or more indirect inspections indicate there may be coating defects.

The suspected areas are classified per NACE SP0502 into these categories:

- **Multiple Severe**–This category is the most severe and should be high priority for exposure and evaluation.
- **Severe**–Severe indications have the highest likelihood of corrosion activity.
- **Moderate**–Moderate indications have possible corrosion activity.
- **Minor**–Minor indications are considered inactive or have the lowest likelihood of corrosion activity.

These classifications will be verified once some of the areas are exposed for direct examination. If indications are more or less severe than the original classifications, then the categories must be adjusted before more digs are performed. These results may also lead to a reassessment of whether ECDA is feasible. However, the industry practice of assigning a minor, moderate, or severe CIS classification based on "dip" alone is not the most effective strategy for identifying the most severe anomalies in a given pipeline region. [Daily, Hodge, 2009]

9.2.3. ECDA–Step Three

In this step, the previous work (steps one and two) are verified with direct examinations. The pipeline is exposed and the coating evaluated for holidays and deterioration. The most severe indications should always be evaluated first to determine if there are immediate external corrosion issues that need to be corrected.

The external examination provides actual information about the quality of the coating and measurement of any external corrosion. An operator exposes areas where coating damage or corrosion is located, continuing until good coating and no corrosion is found. This is one reason older pipelines with poor coating conditions are not candidates for the ECDA process. Any external corrosion must be properly evaluated per the regulatory requirements.

When corrosion is found, an operator must have the proper person(s) available at the dig site to evaluate the coating and any corrosion. The information gathered will help determine the root cause of the coating failure, as well as external corrosion and other issues that are found at the site. This will help ensure future corrosion is mitigated and in determining the corrosion rate and time to next inspection. Some companies now list disbonded CP shielding coatings as the number-one root cause of external corrosion.

When assessing the coating condition, operators identify the type of coating (as much as possible). The adhesion and coating thickness should be checked along with a visual examination for holidays, blisters, disbondment, and damage such as soil stress. These damaged areas should be mapped on the pipe to help determine root cause. Evaluate the pH of the soil surrounding the pipe, especially the pH of areas under any disbonded coating.

The other critical step is to check the pH of any water beneath the disbonded coating to determine if the disbonded coating is a CP-shielding or non-shielding coating system. [Norsworthy 2010] The pH of liquid under a disbonded coating can indicate the effectiveness of the CP current to protect the pipe under the disbondment. A pH of 8.5 or greater usually indicates adequate CP is being provided and corrosion is typically not present, as shown in Figure 9-1. If corrosion is located, it will likely have occurred before adequate CP was applied. With a pH less than 8, active corrosion may be found. This pH is best checked while on-site and as soon as possible after disbondment is exposed. A simple pH paper gives adequate information as to the pH range, but electronic equipment will be more specific.

Figure 9-1. Water under blister on FBE coated pipeline. No corrosion under the blister and water had a pH of 12 indicating that CP was able to protect under the disbonded, non-shielding FBE.

Deposits on the pipe or under the coating should be evaluated to determine if bacteria are present and what the deposits consist of. These deposits can provide valuable insight into root cause analysis.

Unlike FBE, many coatings will shield the CP if disbondments occur. The pipe was not corroded beneath the calcareous deposit, but it was corroded beneath the disbonded heat-shrink sleeve. [Bash, 2010] CP shielding due to loss of adhesion of the pipe coating cannot be reliably identified by the use of any above-ground survey. [Norman, Argent, 2007] We can conclude that the reliability of the ECDA may be weakened by the severe cathodic shielding of disbonded coating. [Kim, Won, Song, 2008]

9.2.4. ECDA—Step Four

Step four requires the accumulation and evaluation of all data gathered from the indirect surveys and evaluation of the exposed coating and pipe. This information helps to determine how effective the ECDA process has been and if it is a viable process for the future on the same or similar pipelines. This step also determines the time until the next survey. Areas for future surveys can be determined with the data obtained during the recent surveys.

9.2.4.1. Exposed Pipe Inspection

Operators can use exposed pipe inspections to determine the condition of the coating and the external surfaces of the pipe. Whether the pipe is exposed from ECDA or ILI data, the same process will be used.

Critical to this process is to have qualified persons on-site to analyze the condition of the exposed coating and pipe. They will be responsible for taking and recording the information for each step of collecting soil and coating samples, taking on-site pH and other required information, and measure and map (Figure 9-2) any corrosion present on the pipeline. They should also take and record the required CP measurements before and after excavation.

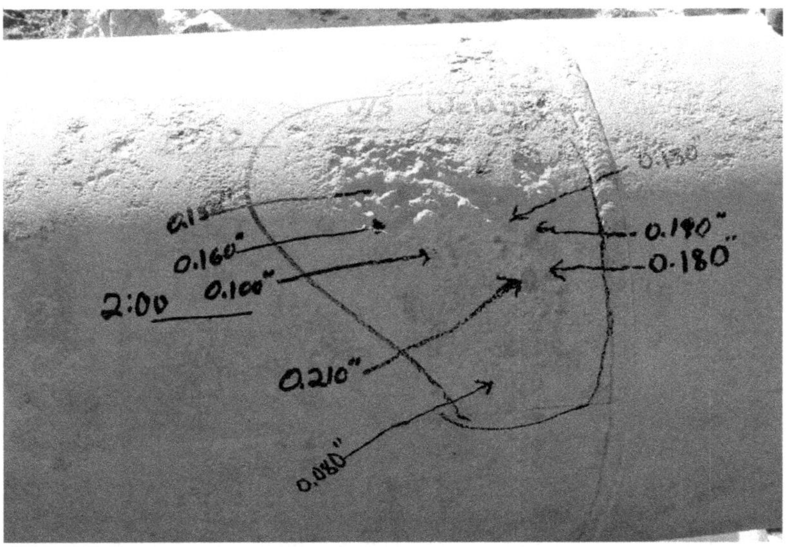

Figure 9-2. Mapping of corrosion on exposed pipeline.

With digital photographic equipment, all phases of the excavation, coating type and condition before and after removal, and pipe condition should be photographed. The location and orientation of any coating damage or corrosion should be indicated in the photos by mapping where it is located on the pipe. These photographs should be recorded for future use. Figure 9-3 shows disbonded solid film-backed tape coating with soil stress.

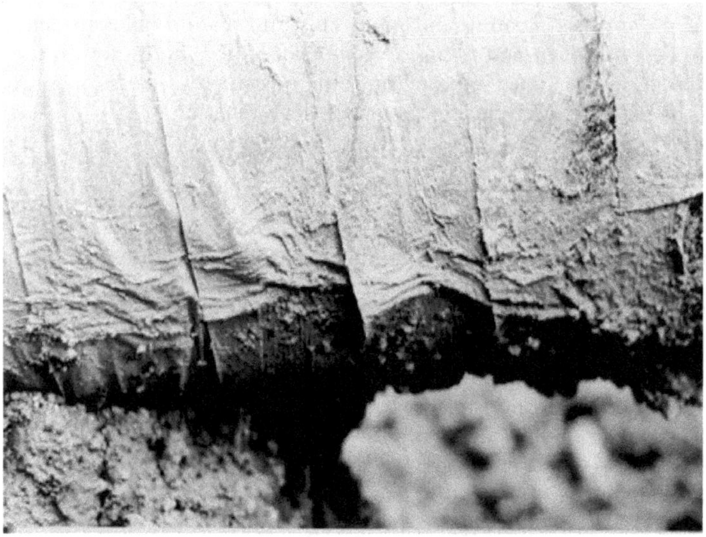

Figure 9-3. Disbonded solid film-backed tape coating.

All safety requirements shall be followed during the excavation, evaluation, recoating, and backfilling to protect the workers, surrounding community, the environment, and pipeline.

9.2.4.2. In-line Inspection

For a direct measurement of what condition the pipe is in and where problems exist, operators can use in-line inspection tools. Several types of ILI tools have been used for over fifty years to determine what is happening with a pipe wall. Several generations and types of ILI tools are now on the market to provide operators with very specific information about the pipe and in some cases the coating type and performance.

None of these tools will find and identify 100% of the defects. The sensitivity of a tool, along with cleanliness of the pipe's internal surfaces, speed of the tool, and other parameters do not allow these tools to be completely accurate. The advancements of technology now provide operators with highly accurate information. Future advancements will only improve the accuracy of these tools.

9.2.4.3. Magnetic Flux Leakage

MFL tools were the first ILI tools developed. These are still very popular because they can be used in both liquid and gas lines since the tools do not require a couplant to perform.

As the MFL moves through the pipe, the axial magnetic flux is induced into the pipe wall between the two poles of the magnet. Corrosion and other pipe-wall defects interrupt the magnetic signal. This interruption is detected and stored in the tool for later evaluation. The number, placement, and sensitivity of the sensors are determined by the manufacturer and is critical to the amount and type of information the tool will provide.

Computers now do most of the interpretation and evaluation of the data to determine defect size and location. With the use of GPS and other spatial electronics, these tools have become much more accurate than before.

9.2.4.4. Ultrasonic Testing

UT tools directly measure the remaining wall thickness as the tool travels through the pipeline. The ultrasonic transducers generate a signal that is perpendicular to the pipe wall. The sound echoes and is received by the transducer. The timing of the return signal allows the tool to determine if there is wall loss from either the external or internal surfaces of the pipe wall.

These transducers must have a liquid couplant between the transducer and the pipe wall for the sound-wave signal to be sent and received. If there is an air space between the transducer and the pipe wall, then the signal will not be sent. For this reason, UT tools are typically used in liquid lines. They can be used in natural gas lines if water is used as a couplant.

9.2.4.5. Electro-magnetic Acoustic Transducer

EMAT tools are part of the latest technology in the ILI field. These tools were initially designed to find various surface cracks in pipelines. As the technology has advanced, some companies have increased the number of sensors and can now locate where coating is either missing or disbonded from the external surfaces of the pipe.

Most of the external corrosion, SCC, and bacteria problems occur on disbonded CP-shielding coatings, this tool now helps the operator find and correct these areas before they become a problem. The continued development and improvements of "Electro-Magnetic Acoustic Transducer" (EMAT) technology to locate and size disbonded coatings without the need to expose the pipeline gives operators economically sound information about their pipeline systems. [Norsworthy, Grillenberger, Brockhaus, Ginten, 2013]

As technology advances, these tools will locate smaller disbondment areas. At this time, some of these tools can also help to identify the type of coating used on the pipeline, providing the operator with critical information. If the coating has been proven to be non-shielding to CP, the operator may want to wait as long as the CP is adequate. If the coating is shielding to CP, then the operator can correct these areas before corrosion or SCC become a problem. This allows the operator to be proactive instead of reactive, as with the other ILI tools.

9.2.4.6. In-line Current Survey Tool (CP Current Measurement)

CPCM is the only method for proactively assessing the effectiveness of a CP system from inside the pipeline. [Janda, 2015] The current survey tool technology continuously measures the voltage drop every 1.0 mm as it travels through the inside of the pipe. The voltage-drop information converts to current, so a detailed report is provided where current is picked up (protection) or discharged (corrosion).

This technology eliminates the need for over-the-line inspections (such as CIS and all the issues with ROW conditions, etc.). In-line current tools are becoming more accepted as the equipment and data interpretation are improved.

The technology also provides information where AC and DC interference issues are located along the pipeline. Bonds, sacrificial anodes, shorts, and other such structural issues are located. The location of CP systems is easily identified, since the direction of current will increase or decrease as the tool approaches or leaves the area of CP system influence. Figure 9-4 shows the in-line current survey tool.

Figure 9-4. In-line current survey tool.

9.3. Coating Condition Testing

Several methods help the operator determine how an external pipeline coating is performing over long periods of time. As coatings age, some properties begin to deteriorate, so more CP is required. In some cases, the pipe is exposed, while in other cases the coating simply starts to lose its original electrical resistance.

9.3.1. Coating Conductance

NACE Standard TM0102 [NACE, 2002b] provides a method for determining the electrical resistance of a coating for a measured distance (usually 1 kilometer or more). The amount of current required from one time to the next is what this test provides. As a coating deteriorates, more current is required. The operator must decide if recoating is required or if more CP will protect the pipe. The type of coating (shielding or non-shielding) will help to determine whether recoating is required or even necessary. Used with ECDA or ILI surveys, this method also provides valuable information for future decisions.

9.3.2. Current Requirement

Current-requirement testing can determine a coating's effectiveness. This test provides the amount of current required for providing adequate protection to the pipeline. The test is performed by applying cathodic protection to a known distance of the pipeline system and measuring the potentials at various points along the pipeline as current is incrementally applied. When a protected level is reached at all the measured points, the amount of current required is used to design the CP system.

9.3.3. Coating Resistance Calculations

Coating-resistance measurements determine the dielectric barrier provided by the coating between the structure and the environment. A good coating will have higher coating resistance than a poor coating.

Specific coating resistance can be determined by multiplying the surface area of the test section by the resistance of the structure to the electrolyte. Doing these tests over intervals of several years help to determine the long-term performance of the coating.

References

Bash, L., "FIELD OBSERVATIONS FROM 56 YEARS OF INVESTIGATING CORROSION DAMAGE ON BURIED STEEL PIPELINES UNDER CATHODIC PROTECTION", CORROSION 2010, Paper 10033

Daily, Steven F., Randy L. Hodge, P.E., INTERPRETATION OF CIS POTENTIAL PROFILE WITH RESPECT TO ECDA METHODOLOGY; CORROSION 2009, Paper 0913

Janda, D., "Capabilities and Advancements in In-line Inspection" Congreso Pemex; November 2015

NACE (2002a) SP0502, Pipeline External Corrosion Direct Assessment Methodology, Houston, TX, USA.

NACE (2002b) TM0102, Measurement of Protective Coating Electrical Conductance on Underground Pipelines, Houston, TX, USA.

NACE (2013) SP0169, Control of External Corrosion on Underground or Submerged metallic Piping Systems, Houston, TX, USA.

Norman, David, Dr Colin Argent, PIPELINE COATINGS, EXTERNAL CORROSION AND DIRECT ASSESSMENT; CORROSION 2007; Paper 07154

Norsworthy, R., Grillenberger, J., Brockhaus, S., Ginten, M.; "EMAT, Pipe Coatings, Corrosion Control and Cathodic Protection Shielding", CORROSION 2013, Paper 0002378

Norsworthy, R, "Finding and determining the cause of external corrosion on cathodically protected pipelines" CORROSION 2010

Norsworthy, R, Grillenberger, J., Heinks, C, Jurgk, M., "Importance of Locating Disbonded Coatings with Electro-Magnetic Acoustic Transducer Technology", Materials Performance 2012

Song, F.M. and N. Sridhar, A SIMPLE METHOD TO ESTIMATE CORROSION RATES AND CP PENETRATION IN COATING-DISBONDED REGIONS ON BURIED PIPELINES; CORROSION 2007; Paper 07155

YoungGeun Kim, DeokSoo Won, HongSeok Song, Il-dong, Sangrok-gu, Kyunggi-do, VALIDATION OF EXTERNAL CORROSION DIRECT ASSESSMENT WITH INLINE INSPECTION IN GAS TRANSMISSION PIPELINE, CORROSION 2008, Paper 08136

Index

D

E

F

G

H

I

P

R

S

T

U

V

W

www.ingramcontent.com/pod-product-compliance
Lightning Source LLC
LaVergne TN
LVHW060032210926
848481LV00013B/3840